Technology for
Global Economic and
Environmental Survival
and Prosperity

Technology for Global Economic and Environmental Survival and Prosperity

Edited by

Behram N. Kursunoglu

Global Foundation, Inc.
Coral Gables, Florida

Stephan L. Mintz

Florida International University
Miami, Florida

and

Arnold Perlmutter

University of Miami
Coral Gables, Florida

Springer Science+Business Media, LLC

Library of Congress Cataloging in Publication Data

Technology for global economic and environmental survival and prosperity / edited by Be-
hram N. Kursunoglu, Stephan L. Mintz, and Arnold Perlmutter.
   p.   cm.
"Proceedings of an international conference on technology for the global economic, environmental survival and prosperity, held November 8–10, 1996, in Miami Beach, Florida"—
T.p. verso.
   Includes bibliographical references (p.   ) and index.
   ISBN 978-1-4613-7732-0   ISBN 978-1-4615-5961-0 (eBook)
   DOI 10.1007/978-1-4615-5961-0
   1. Electric power production—Congresses. 2. Green technology—Congresses. I. Kursuno-
glu, Behram, 1922–   . II. Mintz, Stephan L. III. Perlmutter, Arnold, 1928—
TK1005.T435 1997
333.79'2415—dc21   97-31227
                                                                          CIP

Proceedings of an International Conference on Technology for the Global Economic, Environmental
Survival and Prosperity, held November 8 – 10, 1996, in Miami Beach, Florida

This volume was taken from a series of conferences sponsored by Global Foundation, Inc.,
Coral Gables, Florida

ISBN 978-1-4613-7732-0

© 1997 Springer Science+Business Media New York
Softcover reprint of the hardcover 1st edition 1997
Originally published by Plenum Press, New York in 1997

http://www.plenum.com

PREFACE

The 1996 Conference focused on topics of environmentally attractive technologies for electricity production—renewables, natural gas, and nuclear energy. Recent technology developments were addressed which include creation of more efficient photovoltaic converters for electricity generation; the current and future role of natural gas in meeting global demand for electric power generation; and the status of nuclear energy, its various applications, and the prospects for its future. The Conference agenda, in light of its global economic impact, included comparative discussions of all the above alternative energy sources. The regional choice of energy sources and their impact on the global economy and environment was reviewed.

In addition to the above subjects, but strongly connected with the theme of global energy needs and security, the Conference program contained one session on new needs and directions in higher education: new curricula to cover fundamental global issues on energy, resources, and environment.

The 1997 Conference, which is to be held in Washington, D.C., will further focus on the impact of the use of energy on the environment. For example, could nuclear energy be a major contributor to an energy source mix that can help to alleviate greenhouse gas emissions and global warming problems? To what extent is it true that the use of nuclear energy in rapidly industrializing regions of the world could help to solve regional pollution issues such as acid rain arising from heavy use of fossil fuel, particularly coal? Do the issues of nuclear waste and possible nuclear weapons proliferation present scientific, technological, and political challenges? If all of these problems inhibit the worldwide use of nuclear energy, then the world is facing an important global issue requiring global solutions.

Further, the 1997 Conference will examine the impact of nuclear energy on future regional and global environmental issues on the variety of scenarios that include: (1) competition in deregulated energy environments; (2) constraints levied upon use of fossil energy; (3) possible expansion of nuclear energy into energy sectors beyond generation of electricity (e.g., processed heat, fuel production, etc.).

Behram N. Kursunoglu
Stephan L. Mintz
Arnold Perlmutter
Coral Gables, Florida
February 1997

ABOUT THE GLOBAL FOUNDATION, INC.

The Global Foundation, Inc., which was established in 1977, utilizes the world's most important resource... people. The Foundation consists of great senior men and women of science and learning, and of outstanding achievers and entrepreneurs from industry, governments, and international organizations, along with promising and enthusiastic young people. These people form a unique and distinguished interdisciplinary entity, and the Foundation is dedicated to assembling all the resources necessary for them to work together. The distinguished senior component of the Foundation transmits its expertise and accumulated experience, knowledge, and wisdom to the younger membership on important global issues and frontier problems in science.

Our work, therefore, is a common effort, employing the ideas of creative thinkers with a wide range of experience and viewpoints.

GLOBAL FOUNDATION'S RECENT CONFERENCE PROCEEDINGS

Making the Market Right for the Efficient Use of Energy
Edited by: Behram N. Kursunoglu
Nova Science Publishers, Inc., New York, 1992

Unified Symmetry in the Small and in the Large
Edited by: Behram N. Kursunoglu, and Arnold Perlmutter
Nova Science Publishers, Inc., New York, 1993

Unified Symmetry in the Small and in the Large—1
Edited by: Behram N. Kursunoglu, Stephan Mintz, and Arnold Perlmutter
Plenum Press, 1994

Unified Symmetry in the Small and in the Large—2
Edited by: Behram N. Kursunoglu, Stephan Mintz, and Arnold Perlmutter
Plenum Press, 1995

Global Energy Demand in Transition: The New Role of Electricity
Edited by: Behram N. Kursunoglu, Stephan Mintz, and Arnold Perlmutter
Plenum Press, 1996

Economics and Politics of Energy
Edited by: Behram N. Kursunoglu, Stephan Mintz, and Arnold Perlmutter
Plenum Press, 1996

Neutrino Mass, Dark Matter, Gravitational Waves, Condensation of Atoms and Mono-
poles, Light Cone Quantization
Edited by: Behram N. Kursunoglu, Stephan Mintz, and Arnold Perlmutter
Plenum Press, 1996

Technology for the Global Economic, Environmental Survival and Prosperity
Edited by: Behram N. Kursunoglu, Stephan Mintz, and Arnold Perlmutter
Plenum Press, 1997

25th Coral Gables Conference on High Energy Physics and Cosmology
Edited by: Behram N. Kursunoglu, Stephan Mintz, and Arnold Perlmutter
Plenum Press, 1997

CONTRIBUTING CO-SPONSORS OF THE GLOBAL FOUNDATION CONFERENCES

Electric Power Research Institute, *Palo Alto, California*

Gas Research Institute, *Washington, DC*

General Electric Company, *San Jose, California*

Northrop Grumman Aerospace Company, *Bethpage, New York*

Martin Marietta Astronautics Group, *Denver, Colorado*

Black and Veatch Company, *Kansas City, Missouri*

Bechtel Power Corporation, *Gaithersburg, Maryland*

ABB Combustion Engineering, *Windsor, Connecticut*

BellSouth Corporation, *Atlanta, Georgia*

National Science Foundation

United States Department of Energy

International Conference on

TECHNOLOGY FOR THE GLOBAL ECONOMIC, ENVIRONMENTAL SURVIVAL AND PROSPERITY

(19th in a Series of Conferences on Energy since 1974)
November 8–10, 1996
Eden Roc Resort and Spa
Cotillion Ballroom
Miami Beach, Florida

THURSDAY, November 7, 1996

6:00-9:00 PM   Registration

FRIDAY, November 8, 1996

8:30 AM   **Session I: Global Energy Demand and Energy Security Projections in the Coming Century: Global Economy, Environment, Resources, and Population Issues**

Opening Remarks: **Behram N. Kursunoglu**, Global Foundation, Inc.
"About this Conference"

Moderator: **Richard Kennedy**, Washington, D.C.

Keynote Dissertator for NEA: **Evelyne Bertel**, Paris, France
"What Might Be the Future of Nuclear Energy?"

Scientific Keynote Address: **William F. Martin**, Washington Policy and Analysis, Washington, D.C.
(former Deputy Secretary, US Department of Energy)
"Energy Security in a Global Context"

Keynote Address: **Richard Wilson**, Harvard University
"Closing the Nuclear Fuel Cycle: Reaching a Consensus"

Annotators: **Edward Bacinich**, Alpha Omega Research Foundation, Palm Beach, Florida
Sheldon Freid, Bechtel Nevada
Joseph D. Lehman, Martin Marietta Astronautics Group, Denver, Colorado

Session Organizer: **Behram N. Kursunoglu**

10:00 AM   Coffee Break

10:15 AM	Session I:	Continues
11:00 AM	Session II:	Relevance of International Consensus Policies on Alternative National Energy Strategies as Related to the Regional Needs, Environmental Impact, and Geopolitical Trends

Moderator: **C. Pierre Zaleski**, Université Paris, Dauphin
Dissertators: **Dan E. Arvizu**, Sandia National Laboratories, Albuquerque, New Mexico
"Energy Technology for Sustainable Development"
Jean-Marie Bourdaire, International Energy Agency, Paris
"The Energy Dimension of Climate Change"
Roger Gale, Washington International Energy Group, Ltd., Washington D.C.
"Impact of Electric Utility Competition on Geopolitics"
Alan Pasternak, Lawrence Livermore National Laboratory, California
"Electricity Needs in Developing Countries and Implications for Global Energy Resources"
William Varoquaux, Electricité de France, Paris
"Europe's Energy Trade-Offs: Making Policy Choices in an Imperfect World"
Annotator: **Mariano Bauer**, University of Mexico
Session Organizers: **Session Dissertators**

12:30 PM	Lunch Break	
1:30 PM	Session III:	The Non-proliferation Landscape: 1996 and Beyond—United Nations Comprehensive Test Ban Treaty

Moderator: **Jean Couture**, Paris, France
Dissertators: **Charles Yulish**, U.S. Enrichment Corporation, Maryland
"Status Report on the Russian HEU Megatons to Megawatts Contract"
Richard Kennedy
"General Remarks on Non-proliferation"
Annotator: **William Sutcliffe**, Lawrence Livermore National Laboratory, California
Session Organizer: **Edward D. Arthur**, Los Alamos National Laboratory, New Mexico

2:30 PM	Coffee Break	
2:45 PM	Session IV:	Education on Energy and Global Issues

Gerald Clark, The Uranium Institute, London
Dissertators: **Gerald Clark**
"Educating the Public About Nuclear Energy Issues"
Raymond W. Durante, Durante Associates, Washington, D.C.
"The Eagle Alliance: Grassroots Program to Revitalize Nuclear Science and Technology in America"
Behram N. Kursunoglu
"Nuclear Education on Political and Military Strategies"
Glenn T. Seaborg, Ernest Orlando Lawrence Berkeley National Laboratory, Berkeley, California

"Education for the 21st Century" (presented by Bertram Wolfe)

Annotators: **José G. Martín**, University of Massachusetts, Lowell, Massachusetts

Session Organizer: **Behram N. Kursunoglu**

5:00 PM Conference Adjourns for the Day

5:30–
6:30 PM

Welcoming Cocktails—Lobby Bar Area

SATURDAY, November 9, 1996

8:30 AM

Session V: **Global Initiatives for the Future of Nuclear Energy**

Moderators: **Edward D. Arthur**, Los Alamos National Laboratory, New Mexico

Bertram Wolfe, Monte Sereno, California

Dissertators: **Edward D. Arthur and Richard L. Wagner, Jr.**, Los Alamos National Laboratory, New Mexico

"The Los Alamos Nuclear Vision Project"

Clinton Bastin, U.S. Department of Energy, Washington, D.C.

"Nuclear Technology in the United States—Need for a New Approach: Policies, Programs, and Regulatory Processes"

Peter Beck, Royal Institute of International Affairs, London

"Proposal for a World Commission to Examine the Future of Nuclear Power"

Robert A. Krakowski, Los Alamos National Laboratory, New Mexico

"Long-Term Trade Offs between Nuclear- and Fossil-Fuel Burning"

David Rossin, Electric Power Research Institute (EPRI), San Francisco, California

"The Carter Administration's Decision on Reprocessing in the US"

Ricard L. Wagner, Jr., Los Alamos National Laboratory, New Mexico

"Nuclear Energy and Global Security"

Annotator: **Betrand Vieillard-Baron**, Fromatome, France

Session Organizer: **Edward D. Arthur**, Los Alamos National Laboratory, New Mexico

10:00 AM Coffee Break

10:15 AM

Session V: **Global Initiatives for the Future of Nuclear Energy (continued)**

12:00 Noon Lunch Break

12:00 Noon

Lunch Meeting of the Global Foundation Board of Trustees and Advisors—Topiary Room

1:30 PM

Session VI: **Nuclear Technologies Part A—New Developments and Impacts in Medical and Industrial Applications**

Moderator: **Anthony Favale**, Northrop Grumman Corporation

Dissertators: **D. D. Joel**, Brookhaven National Laboratory

	Dissertators:	**Bernard Appell,*** Electricité de France, Ville Aurbanne, France "The Role of Nuclear Power In Meeting the Energy Needs of Developing Countries" **Louis B. Long**, Southern Nuclear Operating Company, Birmingham, Alabama "Nuclear Power and Its Existence under Deregulation" **Gordon E. Michaels**, Oak Ridge National Laboratory, Tennessee "Status of MOX Reactor Options for Weapons Plutonium Disposition" **Edward Rodwell**, EPRI, San Francisco "Economic Potential of Plutonium" **John J. Taylor** "The International Regulatory Environment"
	Annotators:	**Juan Eibenschutz**
	Session Organizer:	**John J. Taylor**
6:30 PM		Conference Adjourns for the Day
7:30 PM		Conference Banquet—Cotillion Ballroom

SUNDAY, November 10, 1996

9:00 AM	**Session IX:**	**Economics and Politics of an Optimum Energy Demand and Security in Reconciling the Needs of Developed and Developing Countries: Current Status of Oil, Gas, Coal, and Renewables (Solar, Wind, and Fusion), the Environmental Challenges**
	Moderator:	**William F. Martin**, Washington Policy and Analysis, Washington, D.C.
	Round Table	
	Dissertators:	**Jean-Marie Bourdaire**, International Energy Agency, Paris **Richard Wilson**, Harvard University, Boston **Pierre Zaleski**
	Annotators:	**Marcello Alonso**, Florida Institute of Technology **Alan Pasternak**
	Session Organizer:	**Mariano Bauer**
10:30 AM	Coffee Break	
10:45 AM	**Session X:**	**Current Status and Future Global Prospects for Nuclear Energy: Two Scenarios** **1. Can the World Sustain Its Economic and Environmental Needs Without Nuclear Energy?** **2. Can the Use of Nuclear Energy Be Globalized With Built-In Security and Safety Measures?**
	Moderator:	**Pierre Zaleski**
	Round Table	
	Dissertators:	**Clinton Bastin, Gerald Clark, Joseph D. Lehman, John Nuckolls, John J. Taylor, Richard Wagner and Bertram Wolfe**
	Annotators:	**Behram N. Kursunoglu**
12:30 PM		1996 Conference Adjourns

CONTENTS

CHAPTER IV
Global Initiatives for the Future of Nuclear Energy

CHAPTER V
Nuclear Technologies

CHAPTER VI
Key Issues in Nuclear Power: A World Commission on Nuclear Energy

Technology for
Global Economic and
Environmental Survival
and Prosperity

**PRESENTATIONS
ON THE
INTERNATIONAL CONFERENCE
ON TECHNOLOGY FOR THE GLOBAL ECONOMIC,
ENVIRONMENTAL SURVIVAL AND PROSPERITY**

PRESENTATIONS ON THE INTERNATIONAL CONFERENCE ON TECHNOLOGY FOR THE GLOBAL ECONOMIC, ENVIRONMENTAL SURVIVAL AND PROSPERITY

Robert A. Krakowski

Systems Engineering & Integration Group
Los Alamos Scientific Laboratory
Los Alamos, New Mexico 87545, and

José G. Martín

College of Mathematics, Science and Technology
University of Texas at Brownsville
Brownsville, Texas 78520

INTRODUCTION

The Global Foundation, Inc. is a nonprofit organization formed to examine forefront problems in science and technology. The International Conference on Technology for the Global Economic, Environmental Survival and Prosperity, the 19th in the series of Global Foundation Conferences, focused on topics related directly to environmentally attractive technologies available for the long-term generation of electrical power - renewables (solar, wind, ocean-thermal, biomass, etc.), advanced fossil (clean coal, natural gas), and nuclear. The professional interests and expertise of the conferees, and the nature of their presentations, gave the Conference a strong focus on the latter, while recognizing the need for a broad and flexible spectrum of long-term energy options. The Conference reviewed events leading to the present worldwide state of nuclear energy, reviewed a range of non-energy applications of nuclear technologies, examined historical events and decisions leading to the present situation in nuclear energy, and investigated a number of energy futures for industrialized, industrializing, and developing countries. The discussion agenda developed and implemented for the conference focused primarily on comparative assessments of alternative energy sources in a context of a range of possible global futures, particularly as related to energy economics, resources, availability, and security in a world that is moving strongly towards market globalization, energy-service privatization, growing populations desirous of improved living conditions, and looming global environmental

Technology for Global Economic and Environmental Survival and Prosperity
Edited by Kursunoglu *et al.*, Plenum Press, New York, 1997

3

consequences of increased anthropogenic energy conversion. The conferees reviewed higher education needs, and the contribution that academicians can make in providing orientation and coverage on fundamental global energy issues related to resource, equity, environment, security, and stability. A number of presentations emphasized the need for better dissemination of information to the public and decision makers, enabling them to share a clear and balanced perspective of risks and benefits of nuclear energy. Throughout the Conference, attention was drawn primarily to the exploration of the present status and future prospects of nuclear energy, and to related economic and political issues associated with understanding and "optimally" fulfilling global energy demand and security requirements.

THE MEETINGS

SESSION I

GLOBAL ENERGY DEMAND AND ENERGY SECURITY PROJECTIONS IN THE COMING CENTURY - GLOBAL ECONOMY, ENVIRONMENT, RESOURCES, AND POPULATION ISSUES

Moderator: Ambassador Richard Kennedy, 2510 Virginia Avenue, Northwest, Washington, DC 20037

Organizer: Dr. Behram N. Kursunoglu, Global Foundation, Inc., P. O. Box 249055

Keynote Dissertator:
Dr. Evelyne Bertel, Division of Nuclear Power, International Atomic Energy Agency, Wagramer Strasse 5, P. O. Box 100, A-1400 Vienna, Austria

"What Might be the Future of Nuclear Energy?"

While the growth of the nuclear contribution to the overall energy supply the classical s-shaped dependence in its early introduction, that contribution is now beginning to depart from expectations; the market share is now 17% and is expected to decrease to ~15% in the future. The factors driving this divergence include economics; limitations on infrastructure and resource; and issues arising of supply security and diversity issues. Except for the UK (first-of-a-kind Sizewell `B') and the Netherlands, nuclear energy can be shown to be more economically attractive than coal in all NEA countries; depending on the discount rate and the fuel price, the same is true for gas-fired fossil plants. The new factors entering into the nuclear versus fossil comparisons and having an impact on future energy policies are globalization, the market changes induced by deregulation and privatization, and sustainable development goals. Globalization diminishes interest in energy independence as long as energy flows easily across borders. The ease of financial cross-border flow also has an impact financial markets. Deregulation and privatization introduce a certain degree of instability and uncertainty to the markets, have an impact on generating capacity and reserves, and make shorter-term investments more attractive (e.g., the current rush to gas). Sustainable development goals have an impact on the assessment of the relative external costs of atmospheric emissions (either greenhouse gases or radioactive gases) and solid waste disposal (relative to coal ashes, nuclear waste is low in volume but poses a long term hazard). Finally, there is an ongoing reevaluation of the public acceptance and perception of risks of accidents - accidents which, in the nuclear option, have a very low probability but may have grave consequences. The short-term challenge for nuclear energy is the maintenance of the technological infrastructure in a period of low growth and the enhanced competition relative to fossil fuels. The longer-term prospects are more positive for the nuclear option; they derive from its large resource base, mature and demonstrated technologies, a strong potential for future technological progress, and its generally low overall environmental impact.

Scientific Keynote Dissertator:
Dr. William F. Martin, Washington Policy and Analysis, Inc., 1025 Thomas Jefferson Street, NW, Suite 411 West, Washington, DC 20007

"Energy Security in a Global Context"

This keynote scientific presentation was based largely on the recently published Trilateral (U.S., OECD-Europe, and Japan) Commission Report entitled *"Maintaining Energy Security in a Global Context"* (Report #48 to the Trilateral Commission, 1996). This presentation reflects on: a) the nature of the national security threat associated with a growing dependence on Persian Gulf oil; b) projections of both short-term (~15 years) and longer-term (~50 years) patterns of energy supply and demand using the authors' Global Energy Modeling System (GEMS); and c) the possible role to be played by nuclear energy in matching supply with demand on both time frames, while assuring or at least enhancing global and national energy security. The analyses presented *vis a vis* the GEMS analyses are framed in the context of sustainable development scenarios. Achievement of secure and sustainable energy development will require reliance on a variety of energy technologies that include: a) renewables; b) continued increases in energy efficiency and conservation; c) an expanded natural-gas delivery and distribution infrastructure; and d) the maintenance of the nuclear-energy option to the extent needed to ensure cost-effective and safe advanced technologies development and implementation. Short-term challenges for nuclear energy include issues related to the assurance of economic generation, waste disposal, continued safety, and the maintenance of acceptably low proliferation risk in a wide range of international settings. Improved fuel reprocessing and the eventual commercialization of the advanced (fast-spectrum) breeder reactor introduce longer-term challenges. Lastly, a five-pronged strategy for exploring the nuclear option is being proposed here, based on the: a) promotion of an improved understanding of long-term energy supply and demand on a global level; b) continuation and expansion of the comprehensive examination of nuclear futures *vis a vis* the Los Alamos Nuclear Vision Project; c) ongoing discussions with and within the Executive and Legislative branches of the government on long-term energy futures; d) maintenance and nourishment of the nuclear-energy capability and infrastructure (e.g., universities, national laboratories, industries, regulatory bodies, etc.) within the U.S.; and e) continuation and enhancement of consultation and interaction within the international nuclear community through the IEA, IAEA, NEA, and a range of bilateral and trilateral partnerships.

Dr. Richard Wilson, Harvard University, High Energy Physics Lab., Cambridge, MA 02138

A series of crafted rhetorical questions was used to explore the validity of old truisms and to identify potential directions for new solutions to the challenge posed by the title of the presentation. Question #1: "Given that nuclear energy is sustainable (~100,000 years of fuel), creates no greenhouse gases, and creates no air pollution (which is responsible for ~60,000 deaths per year in the U.S.), how well are these three reasons known and how/where does it fit into the national priority?" *Answer: Generally, no one really cares.* Question #2: "Is nuclear a sustainable resource, and what other sustainable options are available?" *Answer: The present administration generally says that other options exist, and nuclear energy is not needed.* Question #3: "What has changed for nuclear since the 1970s? *Answer: Natural gas is cheaper, the uranium supply is larger, and cheap energy does not seem as important.* Question #4: "What is the optimum development of the LMR, and how does that optimum match energy demand with conservation (i.e., efficient use)?" Do reasonable scenarios exist that demand rapid LMR development, as in the 1970s (Fermi's dream)? *Answer: The Ford/Mitre report of 1976-77 and the Carter Decision that followed defined rigidly the proliferation issue in terms that eliminated any reasonable LMR scenario, and that issue remains to be addressed.* Question #5: "What has been done to render an acceptable nuclear energy program vis a vis the concerns expressed in the Ford/Mitre report?" *Answer: Nothing.* Question #6: "Is it true that a "clear and present danger" from weapons plutonium exists, as announced by the Panofsky report (1994?)?" Questions #7: "Given that separated plutonium is a bad idea, how secure is it?; can reprocessing increase the security of plutonium? (e.g., storage of lots of plutonium in one place versus storage of spent-fuel plutonium in many places); how is the public convinced of the security of plutonium in a given form?; what is the cost and does the security issue go beyond cost?; what is the differences between weapons-grade versus reactor-grade plutonium insofar as proliferation value is concerned?; how is safe storage guaranteed?" Question #8: "What are the potential advantages of the IFR, and should the idea be transferred elsewhere?" Question #9: "Can the U.S. nuclear energy program be revived, or will leadership go elsewhere?" Question #10: "What is the cost of plutonium storage versus plutonium burning in MOX?" Question #11: "Should an accelerator be used to breed fuel?; Does the extra flexibility of a subcritical assembly give added proliferation resistance to the nuclear fuel cycle?" Question #12: "What have we learned from past reactor accidents?" Question #13: "Is plutonium really the most toxic material known to man?" Question #14: "To what degree is the NPT a colonialist treaty, and should countries like India and Pakistan work out their own approaches and differences together? Question #15: "What is the most anti-nuclear-energy organization in the U.S.?" *Possible Answers: Ralph Nader?, Friends of the Earth?, the Union of Concerned Scientists?, the Nuclear Regulatory Commission?, the National Resources Defense Council?... the White House!?)*

SESSION II

RELEVANCE OF INTERNATIONAL CONSENSUS POLICIES ON ALTERNATIVE NATIONAL ENERGY STRATEGIES AS RELATED TO THE REGIONAL NEEDS, ENVIRONMENTAL IMPACT, AND GEOPOLITICAL TRENDS

Moderator : Dr. C. Pierre Zaleski, Université de Paris, DauphinPlace du Maréchal de Lattre de Tassigny, 75775 Paris Cedex 16, France

Annotator: Mr. Mariano Bauer, Universidad Nacional Autonoma de Mexico, Apartado Postal 70-172 04510 Mexico, D. F.

Dr. Zaleski opened this session with the following list of key issues: regulation and deregulation; energy security; global climate change (public actions and the impact on energy prices); long-term resources (mainly fossil fuels, with consideration to the time when the impact of depletion is felt); and the perspective from developing countries (related energy models, energy intensity, resources, intercession with respect to greenhouse gases; help from the west). These issues form a recurring theme that in one form or another emerges in many other segments of this Conference.

Dissertators:
Mr. Dan E. Arvizu, Sandia National Laboratory, PO Box 5800 Org. 6200/MSO702 Albuquerque, NM 87185-0702

"Energy Technology for Sustainable Development"

The general desirable attributes of systems for sustainable energy development are that they be affordable and accessible, and that they lead to abundant international commerce. The "top-level" long-term issues underlying the need for this development are related to diminishing oil reserves (20-30 years), increased demand from industrializing countries, and the emerging importance of the impact of energy generation on the global environment. How different technologies can be implemented in different regions will depend on both the needs and the availability of technology, infrastructure, and capital. In the long term future, an advanced hydrogen economy will meet human needs on a sustainable basis. In the meantime, the role of technology is to help fill a window of opportunity in response to decreasing oil reserves. Short- and medium term opportunities for technological contributions include increased extraction efficiency, increased conversion efficiency, improved transmission and distribution efficiency, and increased effectiveness in the finding of new resources. In general, renewable energy sources are a long way from being commercially competitive, but the price of those sources has decreased significantly over the years, as technologies have improved. Sometime in the future, there will be a transition from power generation based on the combustion of methane to generation based on the hydrogen chemical reaction: according to the speaker, fuel cells constitute the key and the basis for future sustainable energy sources.

Monsieur Jean-Marie Bourdaire Director, Long-Term Cooperation and Policy Analysis, International Energy Agency 2, rue Andre Pascal 75775 Paris Cedex 16 France

"The Energy Dimension of Climate Change"

A description of the implications of greenhouse-gas effects on the basis of the approach taken by the IEA was presented. The Rio de Janeiro agreement calls for a return by OECD countries to 1990 emission levels by the year 2000. The Protocol was negotiated by a body with government input (IPCC, an administrative/ministerial/foreign-affairs body,) with little involvement from energy experts. The present situation is defined by an existing infrastructure (e.g., power plants, roads, etc.) and a high level of political momentum associated with that infrastructure and related operations and procedures. The IEA approach reduces energy needs to three essential elements: electricity, mobility (transport), and heating. Correlations of aggregated energy use with time, using expenditure as a surrogate variable, were shown to be highly impervious and rigid to major global events such as oil price shocks. These correlations were viewed as providing a tool for projecting into the future, at least out to the year 2015. Policy options available to control the emission of greenhouse gases in transport include: a) carbon taxes imposed on gasoline; b) a "feebate," tied to fuel efficiency and placed on automobile prices; or c) standardization of automobiles. It was estimated that a 20 \$/tonneC carbon tax would increase the price of a car by ~1% or the capital cost of a coal-fired power plant by ~25%. In imposing these kinds of economic incentives to control CO_2 emissions, it is necessary to position equity as a central issue. A cost-effective and equitable protocol must recognize that there are sectorial differences from country to country, and that there is such thing as a "fossil fuel endowment" - a dependency on past practices which has arisen from the historical development and the resources available to each country.

Dr. Roger Gale President, Washington International Energy Group, Ltd., 1155 21st Street, NW, Suite 202 Washington, DC. 20036

"Impact of Electric Utility Competition on Geopolitics"

The new realities of competition, regionalization/continentalization, demonopolization, and reduced government role through increased privatization in the electric utility industry do not necessarily lead to reduced regulation. In fact, they may have the opposite effect. In the tension between competition and deregulation, movement towards new regulations in the form of price and environmental controls can be expected.

While regulators may claim that they are the agents of change, it is in reality the emergence of new technologies that allow the changes that are taking place. Modular and standardized power plants (whether they are present cogeneration plants or envisioned fuel cells) promise improved deployment and operations by placing more control in the hands of more people. At the same time, an increased awareness of the impact of induced climate change is expected to drive further change. For example, a) replacing existing capital stock with more efficient systems will reduce CO_2 emission per unit of generate energy, but this replacement will place increased demands on funds; b) small operational and system improvements over many distributed systems can lead to large global effects; and c) greenhouse-gas emissions are an efficiency issue in addition to being an environmental one.

The main failure in the implementation of nuclear power, and one that affects safety and security, has been in Eastern Europe. These plants need to be improved.

An important question related to the ongoing process of privatization and demonopolization is what happens in terms of planning and strategizing when a monopoly is broken up. The tendency in other areas - from telecommunications to the airline industry, for example - has been a movement or reconstitution towards a system characterized by increased concentration. The emergence of "mega-projects" and a new centralized system is necessary for the generation of large amounts of capital for investment in large, high-risk projects.

The new systems in the centralized utility industry may focus on small unit sizes (e.g., few-hundred MWe rather than ~1300-MWe units). Which imply no long-term commitment to adverse by-products.

In a globalized economic environment, harmonization of agreements, universalization of standards, and the creation of common environmental regimes are expected to occur. The final result will be a "lean-and-mean" industry. A great opportunity exists for those who can take advantage of it. At the same time, there will be a change in decision making towards more decentralization, moving along the road of new harmonization and new (added, not fewer) regulations.

Alan Pasternak Lawrence Livermore National Laboratory, University of California P.O. Box 808, L-640 Livermore, CA 94551

"Electricity Needs in Developing Countries and Implications for Global Issues Requiring Global Solutions"

[Notes missing]

William Varoquaux Le Sous-Directeur Generation and Transmission Division Electricite' de France 3, rue de Messine 75384 Paris Cedex 08 France

"Europe's Energy Trade-Offs: Making Policy Choices in an Imperfect World"

[Notes missing]

SESSION III

THE NON-PROLIFERATION LANDSCAPE: 1996 AND BEYOND - UNITED NATIONS COMPREHENSIVE TEST BAN TREATY

Moderator: **M. Jean Couture, Paris**. 8 Rue du Jour, 75001 Paris, France

Annotators: **Dr. William Sutcliffe,** University of California Lawrence, Livermore National Laboratory, P. O. Box 808, L-19 Livermore, California 94551

Mr. Charles B. Yulish Vice-President Corporate Communication U.S. Enrichment Corporation, 6903 Rockledge Drive Bethesda, Maryland 20817

"Status Report on the Russian HEU Megatons to Megawatts Contract"

Approximately 500 tonne of Russian HEU (highly enriched uranium), equivalent to ~22,000 nuclear warheads, is under agreement for blending down to ~15,300 tonne of LEU (low enriched uranium) over the next two decades for use in U.S. commercial power plants. Adjusting for variations in Russian production capacity, the following blend-down and purchasing schedule is in force: 6 tonne HEU in 1995; 12 tonne HEU in 1996, and 18 tonne HEU in 1997. Increased HEU quantities are to be determined from 1998 onward. The contracted material has a total value of 12 B$ (8 B$ for enriched product and 4 B$ for contained natural uranium). Including all 1996 shipments, USEC has received 20 shipments of LEU from Russia (~822 warheads); to date it has paid 300 M$ to Russia. A number of startup issues and technical problems have been satisfactorily resolved. The "Mtonne-to-MW" contract has a two-year track record of success; payment issues related to natural-uranium charges have been resolved, and the 1997 contract negotiations (for 18 tonne HEU) are underway.

Mr. Richard Kennedy 2510 Virginia Avenue, Northwest Washington, DC20037

"General Remarks on Non-Proliferation"

Non-proliferation was described as a regime comprised of control, physical security, accounting, international and regional safeguards, and formal treaties (e.g., NPT, CTBT, etc.). The example of the Democratic People's Republic of Korea (DPRK) working with the Korean Peninsula Energy Development Organization (KEDO) was outlined as a case where progress has been made in the face of formidable odds, and one where continued success requires continued support.

The building of two 1-GWe LWRs in the DPRK, along with the transfer of 500 ktonne of uranium, in order to halt the construction and operation of the Magnox-like plutonium production reactors, was deemed especially crucial in turning around what would have been the first withdrawn from the NPT. The precedent such a withdraw would have established and the potential for precipitating the drop-out of other so-called "miscreants" were serious. In exchange for the DPRK's agreement to remain in the NPT and to accept IAEA sanctions, KEDO would design, build, and regulate the two LWRs (Korean ABB BWRs) for the DPRK, and this country would safely assume the operation and management of the plants. North-south politics, the closedness of the DPRK society, the remoteness of the reactor site, and the multilateral KEDO staffing present complicating elements to this deal. Additionally, KEDOs conflicting dual roles (owner and regulator), the transition from owner to supplier, and related protocol issues add to the complexity.

A number of other issues in nonproliferation and arms-controls were addressed. One of these was the upcoming formal meeting of the NPT Committee and the process that remains before the CTBT comes into full force (it must be signed by all nuclear-weapons states). The programs in place for disposition of U.S. and FSU nuclear weapons need not be identical, but the quantities must be.

Lastly, it was pointed out that the efforts to enhance IAEA capabilities are progressing very slowly, particularly in the area of environmental sampling.

SESSION IV

EDUCATION ON ENERGY AND GLOBAL ISSUES

Moderator and First Dissertator: Dr. Gerald Clark,Secretary-General The Uranium Institute, Twelveth Floor, Bowater House 68 Knightsbridge London SWIX FLT UK

Annotators: Dr. Jose G. Martin, University of Texas at Brownsville, 80 Fort Brown Brownsville, Texas 78520

"Educating the Public About Nuclear Energy Issues"

Deregulation and competition are forcing nuclear utilities to make drastic cuts in expenditures, and this serves as background for Dr. Clark's presentation. With a warning that "we neglect education at our peril,", Dr. Clark stresses the importance of retaining an involvement in science education in general, and nuclear science education in particular.

He also shares some basic rules he has learned from consultation with experts in educational matters: 1) whatever is proposed in teaching materials has to be in line with the curriculum; 2) the aim of the efforts should be to open minds rather than to convert; 3) teachers will accept material which is carefully prepared and is not propaganda if the source is clearly declared; and 4) the nuclear industry has much to learn from other industries where stigmatization has not affected *de facto* acceptance.

Dr. Clark suggests that we may need to establish an international association with the mission of creating a code of good practice in nuclear education through the sharing of experiences and success stories. One of its first aims should be to make the basic science more accessible. Another should be to teach the teachers, so that they can fulfill their role as natural allies of scientists and engineers as we break a vicious circle where an anti-science attitudes leads to an anti-nuclear public environment.

Dr. Raymond W. Durante, President Durante Associates, Inc. 1925 North Lynn Street Suite725 Waterview Arlington, Virginia 22209W

"The Eagle Alliance: A Grass Roots Program to Revitalize Nuclear Science and Technology in America"

With the trend in the U.S. to eliminate nuclear science and technology, the risk is great that the nuclear industry will move overseas, and the benefits of domestic sources of both energy and non-energy benefits of "things nuclear" will be diminished or lost in this country. The Nuclear Energy Institute (NEI) is lobbying against these trends from the perspective of the nuclear-energy industry, whereas the American Nuclear Society (ANS) focuses on nuclear data and profession interests. The Eagle Alliance (EA) was formed as a non-lobby, not-for-profit organization with corporate, individual, industrial, academic, etc. memberships. As constituted, it is targeted to fill the gap between the ANS and the NEI.

In its publications, the EA describes itself as "a bold new movement to revitalize nuclear science and technology in America." It intends to provide an educational role that demands responsible answers to claims and counter-claims made and being made of "things nuclear", and will work with the media to propagate information into the present confused state that defines public information, understanding, and related reactions of nuclear technologies.

Its objectives are to solidify public support for nuclear technology; to promote state-level programs to responsibly handle and dispose of low-level radio-active waste material; to simulate national support for programs to systematically eliminate excess weapons plutonium and safely dispose of high-level radioactive waste; to achieve a cohesive, risk-based regulatory regime for all nuclear technological activities; to realize broad application of irradiation technology to improve food safety; and to foster widespread, affordable, and effective nuclear diagnosis and treatment of cancer.

Dr. Behram N. Kursunoglu Global Foundation, Inc, P. O. Box 249055 Coral Gables, Florida 33124-9055

"Education on Global Issues Requiring Global Solutions"

Dr. Kursunoglu reviewed highlights of the history of the ambitious experiment undertaken at the University of Miami to promote education on global issues. Beginning in 1982, a group of pioneers at the University decided on the need to adopt a scientific approach to address global issues, and in particular issues related to energy generation and its impact on global security and equity. A "Distinguished Lecture Series" was an essential element of the University of Miami experiment. Many of the highlights reviewed in this talk were taken from a draft of a forthcoming book on the topic.

Dr. Glenn Seaborg Lawrence Berkeley Laboratory, 1 Cyclotron Road Building 70A/Room 3307 Berkeley, CA 94720

"Education for the 21st Century"

This presentation dealt with the more general problem of rising mediocrity in education in the U.S. and the resultant erosion of our national foundation for scientific literacy. According to this presentation, this educational demise is tantamount to a unilateral "educational disarmament". The skills of our youths are deteriorating at an alarming rate relative to the needs of the society as a whole. U.S. society has evolved through, and emerged from three revolutions: democratization led to the industrial revolution which in turn led to the scientific revolution. The skills and talents that supported and drove research and development in this country have improved products, led to cross fertilization of inquiry and to new products, created new manufacturing methods, produced more complex and automated systems, improved our understanding of quality control, and actually changed basic processes of thinking and reasoning through computers.

While the skills of our youths needed to maintain and advance these achievements are not keeping pace, there are some encouraging signs. Standardized Achievement Test scores of the growing minority groups are improving. The Project 2061 (year that Halley's comet returns) at the E.O. Lawrence Science Museum is improving science education for all youth, and particularly for minority children.

While a broad reassessment of nuclear energy is taking place, moods and fashions change, and we must wait for the right time for a consistent set of standards to emerge. However, the education of small children as to the eventual need for nuclear energy is crucial. *"The attainment of excellence is costly, but mediocrity costs more."*

SESSION V

GLOBAL INITIATIVES FOR THE FUTURE OF NUCLEAR ENERGY

Moderator: **Dr. Edward Arthur**, Los Alamos National Laboratory

Annotator: **Monsieur Betrand Vieillard-Baron,** Tour Framatome, F-92084 Paris-la-Defense Cedex France

Dr. Edward Arthur and Dr. Richard L. Wagner, Jr. Los Alamos National Laboratory, P. O. Box 1663-MS F628 Los Alamos, NM 87545

"The Los Alamos Nuclear Vision Project"

The scope, goals, history, and directions of the Los Alamos Nuclear Vision Project were described. The reduction of the nuclear danger, an understanding of global evolution in "things nuclear", and the identification of strategies necessary to achieve desirable endstates are key elements of this ambitious effort. "Things nuclear" includes nuclear weapons, nonproliferation, nuclear materials, nuclear energy, and related institutional responses and interactions. Among the goals is the avoidance of the large scale threat or use of force, and the realization of the nuclear potential to help meet energy demands and alleviate international and environmental stress.

The perspective of this study is that of the Department of Energy and a national laboratory. The end of the cold war and the simultaneously renewed interest in new directions and constraints related to "things nuclear" was the main initiating event for the Project. The time frame for this ~3-year study is >~ 50 years into the future. This Nuclear Vision Project is in its second year and is comprised of three main operational elements: workshops; supporting analyses; and technical and institutional outreach to other related national and world efforts.

Workshops on nonproliferation, nuclear weapon postures, and the interplay and development of global energy, economics and environmental issues constitute the topics for workshops that have occurred or are presently being planned. The analysis effort (partly describe in a subsequent presentation by Robert Krakowski) is based largely on the construction of analytic scenarios that examine how nuclear energy fits into global, national, and energy security constraints. The outreach component of the Nuclear Vision Project incorporates universities, other national laboratories, and national and international groups concerned with long-term planning and projections of national and global energy needs and related constraints.

Dr. Clinton Bastin NTEU (DOE) 14355 Long Channel Drive Germantown, Maryland 20874U.S.

"Nuclear Technology in the United States - Need for a New Approach: Policies, Programs, and Regulatory Process"

The need for improved worker-manager interactions in nuclear industrial regulatory, and research and development activities was the central theme of this presentation. The potential for highly destructive, high-loss events calls for high-quality and well-conceived controls. Disasters like Chernobyl and the Challenger events were a result of events where managers over-rode the advice of cognizant engineers. Improved labor-management partnerships are needed to improve government efficiency while building needed quality in civilian nuclear operations. Furthermore, the present Nuclear Regulatory Commission confrontational approach to compliance must be replaced by one built on partnering between workers and managers; the replacement will lead to safer and more economic nuclear energy.

The evolution of the Atomic Energy Commission (AEC) into the Department of Energy via the Energy Research and Development Agency intermediate step simply increased bureaucratic folly. An operating corporation to dispel the present operational and managerial immobility is needed. Specifically, the presentation recommended that Congress form a Nuclear Technology Board to review periodically government nuclear programs and strategies for the future.

Other recommendations suggested the formation of a corporate structure led by a presidential-appointed board to deal with strategic gridlock, assure safety, and resolve regulatory differences. An important goal is to eliminate policy "flip-flops" that presently occur with changes in the White House, and to reinstitute the kind of stability brought to things nuclear by the AEC while operating under more corporate-like structure and procedures.

Mr. Peter Beck Royal Institute of International Affairs, Stone House, The Green Frant, Tunbridge Wells Kent, TN3 9DN UK

"Proposal for a World Commission to Examine the Future of Nuclear Power"

It was noted that nuclear energy in the U.S. is in a "bunker;" it is being unfairly attacked, and is hoping to be rescued by some decision to "bring in the tanks" from some ill-defined quarter. Something must be done to get the industry out of this bunker. Since the U.S. is the only remaining superpower, we must be more aware of the impact that the present state of siege in the U.S. has on progress in nuclear power in other parts of the world.

The lay of the present "battle field" is defined essentially by three lobby groups led by: a) environmentalists who firmly believe that nuclear energy is too complex to depend on the stability of humankind for long-term safety and security; powerful, plausible, and respectable arguments are made that the world does not need nuclear power if conservation and renewable energy sources are vigorously pursued; b) nuclear industrialists who make opposite and equally plausible and respectable arguments for nuclear energy to arrive honestly at different results; and c) people who believe in the need for nuclear energy, but who are uncertain about present directions of the technology and who believe changes are needed to make nuclear energy more acceptable.

Generally, the availability of fossil fuels does not justify arguments for the absolute need for nuclear power until ~2050, unless greenhouse-gas accumulations drive serious climate change. At the same time, from present perspectives, the possibility and consequences of global warming are highly uncertain; to deal with this uncertainty, high flexibility is required. To assure flexibility the nuclear-energy option must be maintained - and the proposal for a World Commission to Study the Future of Nuclear Power is the main conclusion from the work summarized by Mr. Beck. [P. Beck, Prospects and Strategies of Nuclear Power, Earthscan Publ., Inc. (1994)]

The World Commission to Study the Future of Nuclear Power would: a) be comprised of a diverse group (environmentalists, nuclear industrialists, government, concerned citizens); and b) would address the economics, safety, nonproliferation, required regulation, and environmental pros and cons of nuclear energy. The primary product and goal of the innovative process introduced by the Commission would be to distill and evaluate a wide range of quantitative and qualitative input to identify pathways by which both environmentalist and nuclear-energy advocates could "get out of their respective bunkers", possibly to the delight and relief of the third body of "lobbyists" listed above. It was emphasized that the choice of this commission must be absolutely neutral to retain and enhance trust in and by those on both sides of the issue.

Dr. Robert A. Krakowski System Engineering & Integration Group Los Alamos National Laboratory, MS F607, TSA-3 Los Alamos, NM 87545

"Long-Term Trade-Offs Between Nuclear- and Fossil-Fuel Burning"

An important aspect of the analytic component of the Los Alamos Nuclear Vision Project described above by Edward Arthur was elaborated in this presentation, and preliminary results from this global energy/economics/environmental (E^3) modeling effort were presented. The Los Alamos group has adapted an existing, well-documented E^3 model originally developed at IEA/ORAU and now used at Pacific Northwest Laboratories - Washington DC to model the economic impact of long-term (~2100), global (nine regions) changes induced by fossil-fuel burning. Incorporation of a nuclear (cost, material flows/inventories, proliferation risk) model that is evaluated "under" the PNL global climate assessment model allows quantitative comparisons of risks associated with both fossil- and nuclear-fuel burning. A business-as-usual (BAU) was reported and compared to a dozen estimates of global and region energy demands. Preliminary results from different scenarios were reported as points-of-departures from this BAU case for a range of plutonium-recycle, nuclear-energy-cost, and carbon (proxy) taxation options.

Mr. A. David Rossin President CISAC c/o Rossin and Associates 24129 Hillview Drive Los Altos Hills, CA 94024

"The Carter Administration's Decision on Reprocessing in the U.S."

When President Jimmy Carter announced in 1977 that the United States would defer indefinitely the reprocessing of spent nuclear reactor fuel, he locked the country into a rigid policy that led to the loss of America's technological lead in nuclear energy. The country has never recovered from the loss of momentum brought about by that policy, and this in turn poses serious downside risks related to national economic interests and global energy security issues.

Dr. Rossin, former Assistant Secretary for Nuclear Energy for the U. S. Department of Energy, gave a provocative presentation on how that policy came about. Although the presentation refers to the the players and the events leading to President Carter's announcement, it focuses on characteristics personality traits of the President himself to find the key to the policy. According to Dr. Rossin, President Carter was truly concerned about nuclear proliferation and he made the reduction of the proliferation threat a personal commitment and a priority administration objective. All technical, political, and diplomatic arguments were to be overridden on the single-minded pursuit of this objective, so that, de facto, the actions called for by Carter's policy would fit well with a political initiative to eliminate nuclear power.

The analytic and political events leading to the April 7, 1977, decision by the Carter Administration: a) to defer U.S. reprocessing of spent nuclear fuel indefinitely; b) to convince other nations to follow the lead of the U.S. in this matter; and c) to insure no discrimination amongst nations in this matter, was painstakingly reviewed in Dr. Rossins' presentation. The issues that shaped the Carter policy remain today and will continue to arise in the coming years, and a broader understanding of the facts and influence that shaped that policy is essential for informed decision in the future.

Dr. Richard L. Wagner, Jr. Los Alamos National Laboratory, 409 12th Street Southwest Suite 310 Washington, D. C. 20024

"Nuclear Energy and Global Security"

A structure for understanding the connectivity between nuclear energy and global security (i.e., the avoidance of a large-scale threat) was described. This structural approach to understanding the benefits (if any) of nuclear energy in stabilizing the international order forms a hierarchy that first ties nuclear energy to (nuclear) fuel supply security, reduced environmental stress (from fossil-fuel recovery and burning), and the filling of energy demand gaps (enough energy). The latter three elements then interconnect to either reduce stresses on the international order or more generally contribute to increased or reduced global security, as defined above. After assessing the relative importance of these interconnections, the overarching considerations and influences of proliferation potentials can be overlaid onto the aforementioned linkages between nuclear energy and global security to provide a structure on which to quantify international security thinking. While this approach to structured thinking is valuable for defining interdependencies, its implementation in the form of a quantitative model represents a future research challenge.

SESSION VIA

NEW DEVELOPMENTS AND IMPACTS IN MEDICAL AND INDUSTRIAL APPLICATIONS

Moderator: Mr. Anthony Favale, Deputy Director, Energy Systems Northrop Grumman Corporation, Mail Stop B29-25 1111 Stewart Avenue Bethpage, NY 11714-3588

Annotator: **Dr. Arnold Perlmutter,** University of Miami, Coral Gables, Florida 33124-5510

Mr. Darrel D. Joel Medical Department, Brookhaven National Laboratory, Upton, NY 11973

"Boron Neutron Capture Therapy on Malignant Brain Tumors at the Brookhaven Medical Research Reactor"

Because of the preferential absorption of certain boron-containing chemicals in malignant brain tumors (a relative absorption of about the ~4-5 to one), enhanced radiation dose to a tumor area through the nuclear reaction $n(^{10}B, ^6Li)alpha$ has led to factors of 4-5 increase in tumor-growth/reversal rates. Furthermore, the use of higher intensities of epithermal neutrons leads to higher thermal neutron fluxes at the tumor site as these higher energy neutrons are thermalized in surrounding healthy tissue. This further reduces the radiation doses delivered to the healthy tissue. Life extension has been achieved; in addition, there has been an improvement in the quality of life of the patients: only one irradiation is necessary, instead of multiple ones, and side effects are significantly reduced. Although statistics on the 15 patients treated so far with Boron Neutron Capture Therapy (BNCT) are not sufficient to draw strong conclusions, preliminary observations appear to be encouraging for seriously ill patients - that is, those with a life expectancy of a few months.

Dr. Joseph J. Sredniawski Northrup Grumman Corporation, 1111 Stewart Avenue Bethpage, NY 11714

"A New Approach to Contraband Detection in Luggage and Cargo"

In the technique described in this presentation, protons produced in a few-milliAmpere Accelerator are directed to a water-cooled rotating target to generate 8-9-MeV gamma rays. The gamma rays are collimated into a nominally flat (conical) disc. Luggage or cargo articles are positioned along the outer radius of the disc source of energetic gamma rays. Nitrogen (^{15}N) and/or carbon (^{13}C) - bearing materials undergo resonance absorption and tomographic techniques can be used nondestructively to provide images of objects.

Since most chemical explosives contain nitrogen and carbon in specific ratios, this sensitive imaging technique provides a unique means to detect these clandestine materials. The technique can to extended with somewhat reduced selectivity to the detection of contraband drugs. Although somewhat bulky, the detector system, consisting of a proton accelerator, target, collimator, shield and detector, can be mounted on a truck to provide some degree of mobility.

For airport applications, the accelerator would be installed below (in a basement) the target/collimator system, and a rotating carrousel would bring luggage/cargo items to a detection ("interrogation") position between the gamma-ray beam and the ring of detectors. "Interrogation" rates as high as 800 items/hr are possible. Through the clever use of geometric attenuation, modest amounts of gamma-ray shielding (few centimeters) are required to protect personnel at the periphery. Lead shielding, thicker than 10 cm, is required at the interrogation site, however.

Although detailed "bottoms-up" cost estimates have not yet been made, an accelerator of this size is expected to cost at least 1 M$; costs for the entire unit are expected to exceed a few M$. The basic attenuation and the bulky-albeit-portable nature of the systems probably precludes field applications involving the detection of plastic land mines.

NUCLEAR TECHNOLOGIES - THE NEXT GENERATION OF REACTORS

Moderator: **Dr. Steven A. Hucik**, General Electric Company, San Jose, CA

Dr. Steven A. Hucik* General Electric Nuclear Energy 175 Curtner Avenue M/C 780 San Jose, CA 95125

Advanced Liquid Metal Breeder Reactor"

This presentation focused on recent successes overseas by General Electric (GE) in marketing (Taiwan and Japan) and building (Japan) Advanced Boiling Water Reactors (BWR's). The twin 1300-MWe ABWRs completed for the Tokyo Electric Power Company were fully designed and fabricated by GE and installed in a record 48 months. The conceptual design for the ABWR began in 1978 and after two years moved into the 1980-85 conceptual design phase. This Nuclear Regulatory Commission (NRC)-approved standardized design is expected to be replicated 13 more times in Japan over the period up to 2004-05; in addition, to two units are being planned for Taiwan. The ABWR represents a series of over a dozen important, but incremental, improvements to the basic BWR design, which include improved materials corrosion behavior, water chemistry, internal water pump designs, control-rod drives, advanced core and fuel designs, etc.

The evolutionary trend for the ABWR is towards increased capacity (1500 MWe) and reduced plant complexity to improve economics while retaining good (passive?) safety characteristics based on proven technology for further.

General Electric retains the "know-how" to buildup the Advanced Liquid Metal Reactor (ALMR) insofar as the basic reactor and associated fuel cycle are concerned, but that system at present is not economically competitive. The general approach for ALMR market penetration is to maintain relatively small (300-MWe) units as the basic module size to be used to scale to larger, more-economic overall plant capacities. This approach is thought to reduce the risk associated with the introduction of these new technologies while taking advantage of known economies of scale, as compared to a direct scaleup to larger unit sizes.

The slowing of the LMR development has provided time to re-evaluate the overall approach to these advanced reactor systems. This period of re-thinking the LMR has already produced considerable reductions in overall plant size compared to systems envisaged through the lens originally provided by the original Clinch River and Super Phoenix approaches.

* presented by **Charles Boardman**

Mr. Juan Eibenschutz Hartman, Subdirector de Distribución, Luz y Fuerza del Centro, Melchor Ocampo 171, Piso 8, Colonia Tlaxpan, 11379 Mexico, D. F.

"The Future of Nuclear Power"

This paper gave a lucid, comprehensive, concise, and insightful coverage of a broad range of issues related to the future implementation of nuclear energy. In the past, government support was fundamental to nuclear energy development, and that support has made possible all of the research and development needed for the safe introduction and implementation of commercial nuclear power.

Entry of nuclear energy into the market place, however, would not have occurred if it were not economic. With the present trends toward globalization, privatization, and forced competition, the energy supply in many parts of the world is being left to the invisible hand of market forces. Nuclear energy generally is unsuited for this situation, and the risk of stranding plant investments exists if the price of electricity from nuclear sources cannot compete with sources generated from cheap fossil fuels that are converted with high efficiency in relatively low-capital plants. Changes in fuel prices, however, could lead to other conversion units being stranded.

The rate payer must pay one way or another for the security of the energy supply. Since electrical energy cannot be stored, erosion in capacity (shortfalls) can be more expensive than over-investment in generation capacity. In this instance, governments will step into the market to either make up capacity deficits directly, or to institute regulations that in one way or the other will assure security of supply; electricity for both industrialized and industrializing countries is too central to overall economic prosperity and security.

Generally, Independent Power Producers (IPPs) will avoid electric energy systems like nuclear or hydroelectric plants; high capital investment and strong regulatory constraints are an anathema to the short-term profit motives of IPPs compared to the longer-term requirements associated with maintaining market share and long-term fuel supply. While gas-fired conversion is at present cheaper, methane contributes to growing concerns of greenhouse-gas induced climate change, both in terms of the increased CO_2 generation (albeit, at rates that are ~60% that of coal) and in terms of the direct release to the atmosphere as usage continues to increase.

There are technical solutions to the disposal of the radioactive waste from nuclear plants, but dealing with well-known problems on a subjective basis has been much more difficult. Generally, it was felt that nuclear energy has done a better job in internalizing its costs than most of the alternatives. For example, what would be the cost of hydroelectric power if we were to charge a hydroelectric station for site restoration costs?

Diversity in the global energy supply is crucial. While segments of the public feel that a portfolio consisting of solar, biomass, and ocean-thermal plants can replace the burning of fossil and nuclear fuels, a future of diverse energy supply and environmental constraints without a contribution from nuclear energy will be difficult indeed.

Dr. Edward Teller, Dr. Lowell Wood, and Dr. John Nuckolls* Associate Director Emeritus Lawrence Livermore National Laboratory, University of California P. O. Box 808 Livermore, CA 94550

"A Positive Approach to Nuclear Reactors"

The main problems of radiation, safety, cost, proliferation, and fuel breeding associated with the long-term implementation of nuclear power were addressed in this novel approach to nuclear power. In this approach, a ~1-m- radius cylinder ("stick") of nuclear fuel of length ~ 10 m would be placed 100 m under ground. A fissioning "wave" would be initiated by the withdrawall of a liquid ^{6}Li control medium at one end of the (6-8 tonne/m^3) fissioning "stick" of U/Th fuel. The fission wave would propagate unattended along the "fission stick" at a rate of a few cm per month; and approximately 50 years would be required for the wave to propagate the full length of the ~10-m "fission stick".

Heat would be extracted and delivered 100-m to the surface by helium coolant for conversion at high efficiency to electricity. Automatic control of the burn/propagation rate would be provided by a self-motivated liquid-^{6}Li stream that is moved thermal-hydraulically into and out of the "fission stick" by temperature differences between the "fission stick" per se and the exit helium coolant; ~1000 such ^{6}Li control loops are envisaged to provide the redundancy necessary to assure absolute unmanned control of power generation in the fission stick.

Essential to the slow propagation and efficient burning of the "fission stick" is the breeding of ^{233}U in advance of the slow fission-driven ("glacial") deflagration and the requisite (materials-dependent) neutron mean-free-path. After about 50 years of unattended fission, the burned-out fission stick would be left securely *in terra* as a kind of final disposition that would be unapproachable by future proliferators.

As the authors of the idea acknowledge, the "fission stick" is a design of a conceptual power plant performed by physicists, and is in need of serious engineering scrutiny.

* presentor

SESSION VII

CURRENT STATUS OF FUSION ENERGY

Moderator: Dr. John Nuckolls, Lawrence Livermore National Laboratory, P. O.Box 808, L-1 Livermore, CA 94551

Annotators: Dr. R. A. Krakowski, Los Alamos National Laboratory, MS F607, TSA-3 Los Alamos, New Mexico 87545

Dr. Emilio Panarella Editor Physics Essays Advanced Laser and Fusion Technology Inc. 189 Deveault Street, No. 7 Hull, P. Q. J8Z 1S7 Canada

"Fusion as a Technology of the Global Economic, Environmental Survival and Prosperity: Problems with the Present Approach"

A serious problem with the present direction of fusion energy research was expressed and provided the focus of this paper. These problems were documented at a recent (1994) symposium in Washington organized by the author and reported in a recent issue of the

Journal of Fusion Energy; a follow-on symposium is planned for Washington DC in 1997. While fusion has been and continues to be viewed as a long-term R&D program of glacial migration through experiment and computation towards a usable commercial end-product, approaches that portend much shorter development times can be envisaged. These approaches were describe in the aforementioned symposia. Scientific progress in containing burning deuterium-tritium plasmas has been great, and economic feasibility is replacing scientific feasibility as a measure of success for fusion energy. The present tokamak (magnetic fusion energy, MFE) and laser (inertial fusion energy, IFE) approaches are not the best route to making a fusion power plant that a utility would want to buy. The present ITER (International Thermonuclear Engineering Reactor) tokamak MFE approach may be acceptable for a rich world, but its complexity, size (extremely low engineering power density, even compared to HTGRs), technological stress. etc. are inappropriate for the real world of limited resources, risk-adverseness, and dwindling time lines.

The recent arguments for and against the ITER approach published in *Physics Today* (June 1996) were reviewed (e.g., cost, size, bigness of development step, lost development and commercialization flexibility, narrowed focus too early, risk of partnership pullout, risk of single-point failure, etc., versus the need to focus R&D to study difficult problems of non-linear physics in a full-size and integrated device that provides a cheap buy-in for the U.S.).

In addition, scientific arguments against the ITER, arguments were made to the effect that non-linear physics is best (i.e., more cheaply) understood in small experiments supported by computation. Existing (tokamak) machines are already huge in size and budget, and the cheap (5%) buy-in for the U.S. would still be expensive if (other) partnership pullouts occur and the program collapses. Though this "cheap buy-in", limited U.S. resources would be misdirected away from more fruitful approaches to fusion energy.

The presenter argued that the inertia of international agreements may stall scientific progress. Generally, the present research and development timelines push fusion far into the future.

Ms. Valerie Debski Westinghouse Electric Corporation, 4400 Alfay Trail Orlando, FL32826

"Westinghouse Status of Coal and Biomass Fueled Combustion Turbine Systems for Power Generation"

After sharing data on the present worldwide resources and usage of coal, recent progress in developing clean-coal technologies at Westinghouse was reviewed in a very different and enlightening presentation.

Progress centers around component development for systems based on pressurized coal fluidized bed (PCFB) systems operated with and without topping cycles. The latter require equipment component advances (syngas combusters and hot filters) that allow the production of syngas and its combustion at increased temperature for use in an advanced gas turbine (GTCC). Generally, orders for both PCFBs and GTCCs are increasing (mainly overseas). A smaller plant footprint and lower cost are the main reasons for the interest in the GTCC and PCFB cycles.

Westinghouse started syngas-fired combustion turbine development in the 1970s. Enabling technologies are: a) the hot-gas filter (ceramic "candle" elements that remove erosive particulates and corrosive sulfur prior to entrance to the combustion turbine); and b) a special topping combuster for syngas (a multiannular swirl-type burner that receives a mixture of powder-injected coal and limestone).

While the Westinghouse clean-coal program is centered in Lakeland Florida, a biomass unit operates in Maui Hawaii on sugar cane (~100 tonne/day, 20 MW). A second biomass unit in Minnesota operates on alfalfa stems and at a capacity of 75 MW produces low-caloric gas in a combined cycle unit. The ability of the latter unit to operate with diverse fuels is being demonstrated.

Central to both the implementation of clean-coal technologies and fuel-diverse biomass converters is the advanced combustion turbine technology. Westinghouse is now working on the third generation of this enabling technology, with conversion efficiencies increasing with capacity fromm ~40% at ~100 MWe to ~60% at 400 MWe.

SESSION VIII

KEY FUTURE ISSUES IN NUCLEAR POWER: NEEDS OF DEVELOPING COUNTRIES; INTERNATIONAL MONITORED RETRIEVABLE STORAGE; REGULATORY ENVIRONMENTS; NUCLEAR POWER IN DEGREGULATED GENERATION ENVIRONMENT

Moderator: **Dr. John Taylor,** Electric Power Research Institute, P. O. Box 10412 Palo Alto, CA 94303

Annotators: **Mr. Juan Eibenschutz Hartman,** Luz y fuerza del Centro, Melchor Ocampo 171, Piso 8 Col. Tlaxpana 11379 Mexico, D. F.

This session was prefaced with remarks by the Moderator indicating that the key issue for nuclear energy is to maintain the option for future use. It was noted that the Advanced Light Water Reactor was pushed by EPRI to make that option available when needed. The Advanced Boiling Water Reactor and the System 80+ have been reported to and approved by the NRC to demonstrate the practicability of "one-stop" licensing. Lower power output was needed for passive safety while finding ways to counter the cost of lower-capacity units. The NRC approval has helped General Electric move into the Japanese and Taiwanese nuclear-energy markets.

Dr. Bernard Appell; Electricité de France, Direction de L'Equipment 12-14 Avenue Dutrievoz, 69628 Ville Aurbanne, Cedex, France

"The Role of Nuclear Power in Meeting the Energy Needs of Developing Countries"

The role of nuclear power in developing countries is made complex by issues related to population growth, economic risk, and infrastructure inadequacy. The focus of this dissertation is Asia and Latin America, with the developing countries being divided into: a) newly industrialized countries (Singapore, Taiwan, South Korea); b) fastly industrializing countries (Argentina, Brazil, Mexico, China); and c) yet-to-be industrialized countries. In the 1990-2020 time frame, energy consumption is expected to increase by approximately 10% in the developed countries and by approximately 200% in the developing countries, suggesting a cumulative world energy consumption of 300-400 Gtoe by 2020, increasing to ~1300 Gtoe by the year 2100. CO_2 emission will increase by 50-100%, depending on the degree to which the world uranium resource of ~17 Mtonne is utilized.

The following minimum constraints were suggested as necessary for a developing country to utilize nuclear energy: a) be a signatory of the Non-Proliferation Treaty; b) acceptance of IAEA inspections; and c) implementation of a national security organization. To assure that the needed technical "know-how" and infrastructure is adequate for implementation of nuclear energy in a given developing country, a) the entry level of development should assure mastery of all required technical aspects; b) a plan for long-term development of the technical infrastructure should be in place; and c) the electrical network capacity should be at least ten times the connected opacity.

The prospects of nuclear energy in Asia (17% of world energy resource, 20% of world energy consumption, electrical energy growth at ~7%, nuclear power amounts to ~10% of the world usage) and Latin America (which consumes 6% of the world energy, possesses 12% of the proven oil resources, 6% of proven gas resources, 2% of coal resources, with 37% electricity from fossil, 19% from hydroelectric, and 2% from nuclear) were assessed by country (China, Taiwan, Indonesia, Argentina, Brazil, and Mexico). If met by fossil fuels, the overall ~450% increase in electrical power generation foreseen by the year 2020 poses major CO_2 and energy-dependence problems. It was concluded that only nuclear energy can satisfy future needs beyond the year 2020.

Dr. Louis B. Long Southern Nuclear Operating Company, 42 Inverness Center Parkway Birmingham, Alabama 35242

"Nuclear Power and Its Existence Under Deregulation"

The Southern Nuclear Operating Company is the largest nuclear holding company in the U.S. (it generates some 30 Gwe.) Like many similar corporations facing the low-price markets being created by deregulation in this country, the company is going overseas, where markets are more attractive from the perspective of a firm that buys and deals in electric generating companies.

The strong move towards deregulation to achieve cheaper electrical energy (GTCC generates energy at 2.7 c/kWeh) is being driven by the consumer. Nuclear energy, with its high investment cost and low production cost, will be left with significant stranded costs (unpaid and unpayable capital charges) upon deregulation-drive decreases in cost of electricity. (Write-offs, exit fees, or accelerated depreciations are some of the main ways in which stranded costs can be subsumed.)

Although capacity factors for nuclear plants have increased through shortened outages, plants with high production costs must be shut down. Beyond the year 2000, a production cost of 1.4 c/kWeh is given as a goal, with this goal being met by a combination of increased plant performance goals, capacity upgrades, and reduced operations costs (of 18%). Base-load capacity additions are not needed until the year 2009.

Dr. Gordon E. Michaels Oak Ridge National Laboratory P. O. Box 2009 Building 9102-1, MS-8038 Oak Ridge, Tennessee 37831-8038

"Status of MOX Reactor Options for Weapons Plutonium"

Participants in the MOX options for disposition of weapons-released plutonium are LANL, ORNL (lead), and a range of NSSS vendors. This option requires a pit processing plant (1 B$, 10-year period), a MOX fabrication plant (100 tonne/yr, in the U.S. or possibly in Europe), and LWRs (either in the U.S. or Europe). Four reactor options based (optimistically) on a 100% MOX core loading are CANDUs, Evolutionary LWRs, Advanced LWRs, or a Partially Complete LWR. The reactor option is the cheapest (~2 B$) of the weapons- plutonium disposition options (e.g., immobilization on glass and disposition in a repository, MOX + immobilization hybrid, and deep borehole), and is based on the best known/demonstrated technology. These advantages, however, are not well advertised in comparison to the lesser advantages being promulgated by advocates of other disposition options.

Dr. Edward Rodwell Electric Power Research Institute, 3412 Hillview Avenue Palo Alto, CA 94303

"Economic Potential of Plutonium"

The results of a recent EPRI *study (Electric Power Research Institute report TR-106072, February, 1996)* were reported. The basis of this study is the supposition that the global use of nuclear energy in the next 50 years will be sufficient to deplete the uranium resource to such an extent to make plutonium presently being "stored" in spent fuel economically attractive to low-enriched uranium as a replacement fuel. The main inputs and uncertainties addressed in the EPRI study include: a) the cost of spent-fuel reprocessing; b) the relative unit costs of LWRs versus LMRs; c) the relationship between the cumulative uranium usage and the unit cost of mined/milled/converted uranium ore; and d) the global growth of nuclear energy. If uncertainties in each of these four key inputs/variables are varied separately over reasonable ranges, the middle bound on the time when plutonium usage become economically competitive with ^{235}U is in the range 2025-2050; stacking together of any two out-bound uncertainties, however, push this introduction date to beyond the year 2050.

The optimum starting date for commercial LMR deployment is highly uncertain, with roughly an even chance that it will antecede the time when the LMR can be made ready for such a deployment. Although obvious, it bears restating that economic disincentives exist that preclude the processing of plutonium and the fabrication of MOX fuels until a specific intended use is identified; the best place to store plutonium is as spent fuel until it is needed. Generally, the present program for disposing of spent fuel (in repositories) is unlikely to create a significant constraint on the implementation of the economically optimum LMR scenario. The greatest incentive for early LMR introduction is to reduce the costs of reprocessing and the LMR per se, while continued global depletion of the world uranium resource sustains a gradually growing incentive for LMR introduction.

SESSION IX

ECONOMICS AND POLITICS OF AN OPTIMUM ENERGY DEMAND AND SECURITY IN RECONCILING THE NEEDS OF DEVELOPED AND DEVELOPING COUNTRIES -CURRENT STATUS OF OIL, GAS, COAL, AND RENEWABLES (SOLAR, WIND, AND FUSION), THE ENVIRONMENTAL CHALLENGES

Moderator: **Dr. C. Pierre** Zaleski, Universite' Paris, Dauphine)

Annotator: **Dr. Marcello Alonso,** Melbourne Beach, Florida 32951

Dr. C. Pierre Zaleski Universite' Paris, Dauphine Place du Marechal de Lattre de Tassigny 75775 Paris Cedex 16 France

"Ideas for Discussion"

In formulating ideas for discussion, two primary problems dealing with energy demand and security must be recognized: a) environmental constraints related primarily to global climate change; and b) the exhaustion of fossil fuels. Gas reserves are believed to outlast oil reserves, while coal is expected to outlast them all. Regional depletion of resources, however, will occur sooner and will lead to geopolitical instabilities. Since market forces cannot anticipate or deal with these two key problems (because of short-term view or perspective, and because of inertial effects), governmental actions are needed and will occur.

The questions then must revolve around: a) the kinds of actions; b) the place where such actions should be applied; and c) when such actions should be made. If too strong and/or poorly timed or applied, government actions can upset economies and markets, and could disrupt existing facilities or attempts at corrective indigenous actions.

In setting the stage for a plan of action, it is convenient to divide the world into four regions in order to develop a sense as to where action will be needed:
- industrialized countries: stable growth and energy consumption;
- transition countries (FSU): eventually hope to move to industrialized category;
- low-wage countries with access to technology (China, Indonesia): heavy contributor to CO_2 emission, heavy use of fossil fuels, > 1 billion people; strong potential for growth, increased living standards, energy consumption, etc.

Other developing countries are not a major contributor to the problems and needs listed above. The question of when corrective action should be taken is dictated primarily by two time constants: a) that for CO_2 impact; and b) that for energy resource depletion. If step changes are to be avoided to minimize destabilizing effects on economies, recognizing that the full impact of any change or action will take decades, and recognizing that the CO_2 impact has 50-100 year time constants, related actions must begin now.

With respect of fossil-fuel depletion, regional deprivation and related instabilities represent the main problem; in this context, 20-30 years may be adequate, but one begins to worry for times latter than ~ 40 years The kinds of corrective actions can be grouped according to the four regions listed above. Moderate greenhouse-gas taxes could be applied to the first group of countries, with different tax rates being assessed for electricity, transportation, etc. end uses. This tax could be used to develop new technologies. The use of incentives is suggested for the third group of countries listed above to induce the use of cleaner fossil fuels or the use of non-fossil fuels (e.g., guaranteed loans for nuclear or renewable energy sources, subsidize high-technology exports, etc.). A number of incentives for helping the third group of countries can be identified: a) more efficient use of finances (they are building new plant, the first group of countries are not); b) subsidies help the first group of countries in developing high technology systems and to export them; c) mutual market development is fostered; and d) the right that each country has the same greenhouse-gas emissions is assured.

Monsieur Jean-Marie Bourdaire Director, Long-Term Cooperation and Policy Analysis International Energy Agency, 2 rue Andre Pascal 75775 Paris Cedex 16 France

"Energy Demand, Nuclear Energy, and the World Market"

Because of unfavorable past experiences, the IEA is not well disposed to subsidies, and then only if subsidies are explicit and directed. The suggestions made by the previous speaker must be cognizant of these historical constraints. Given that demand equals the sum of electricity, mobility, and heat, and given strong historical evidence that the demand for each is little affected by major calamities such as oil shocks, little flexibility exists to manipulate the demand side except for truly major dislocations.

Hence, demand is well defined (at least over the next 2-3 decades), and the main question is how best to fulfill that demand. Total demand (linked to gross domestic product) is first supplied by "unavoidable" energy sources (hydroelectric, nuclear, fossil outside of OPEC). Oil from OPEC, or more specifically, oil from the Persian Gulf states, is the "swing fuel". Furthermore, Saudi Arabia is the swing member of OPEC - the swing supplier to meet the aforementioned demand gap. The world hydrocarbon reserves were listed as follows

country/region	percentage	oil	gas
Middle East		64.9	32.4
FSU		5.8	40.4
OECD		10.3	10.0
non-OECD		19.1	17.2

Europe depends on a third of its gas and oil from outside source, and this dependence will shift to one half by the year 2010. The introduction of nuclear power was not driven by the market; nuclear power was introduced when energy was relatively cheap - it was started as an artificial program for a number of non-economic (short-term) reasons. Hence, nuclear energy would be supported by market forces only if environmental forces rule; otherwise, strong government regulation is required if nuclear energy is to make a contribution.

Dr. Richard Wilson; Harvard University, High Energy Physics Lab Cambridge, MA 02138

"Questions of Contempory View on Nuclear Energy and Developing Countries"

Three common assumptions about energy in developing countries were considered, and their applicability/validity investigated. First, it is often stated that nuclear energy is not appropriate for developing countries. However, a "before and after" picture of South Korea and of Japan would tend to refute this claim. A similar conclusion would be reached in another context when Egypt's ability to run the Suez Canal was questioned in the 1950s, and when Egypt hired the best experts (from Holland) to transfer that expertise to local personnel. Generally, it was claimed that civilian nuclear facilities can be run by a cadre of hired experts who can teach local people their skills.

The second assumption is that the free market will "take care of it all." Unrestrained capitalism, however, will ignore and destroy the under- or undeveloped. If free-market forces reign, the gap between the rich and the poor will grow. Advice and help is neeed to assure against this impersonal market, which is a concern that grows as the forces of privatization in the free market loom large. Generally, market forces will not help the developing countries, but will only deplete their resources to the general benefit of the exploiting developed country. The developed countries, in the era of globalization of both trade and environmental effects, must look after the health of its neighbors, else the latter will catch a disease and pass it on to the former.

Thirdly, it is often claimed that small is beautiful and that the renewables (solar, biomass, wind, etc.) are better suited for the developing countries. The renewables will work well only if a strong network exists, and such networks are weak or non-existent in the developing countries; generally, the renewable energy sources are probably easier to implement in the industrialized countries. Furthermore, developing countries lack skilled people below the thin cadre of trained bureaucracy that generally runs the country; decentralized systems associated with renewable energy sources need a wide base of skilled people, and this base does not exist in developing countries. Finally, habits and cultures must be changed in implementing renewable energy sources, and these changes can be difficult.

Lastly, the U.S. has more than 400 years' worth of reserves of coal and it has the potential for methanol and/or hydrogen for transportation. Therefore, it can be argued tht it has no short-term intrinsic need for nuclear energy (the need to retain a place at the global nuclear table, however, cannot be overstated). Furthermore, distribution of electrical energy to the household end-user can easily double the cost of electricity, so distribution is an important part of the cost, and diffuse renewables may have an advantage in that regard. While carbon, particulate, and security (for use of energy derived from OPEC countries) taxes are options for controlling greenhouse-gas emissions and the impact of climate change, probably the cheapest route to reducing CO_2 emission is to reduce these emissions in China and then to get credit for it.

Generally (Peter Beck), nuclear energy cannot be left to market forces, since market forces are short-term and OPEC has much cheaper oil that presently is shut in and could be released at any time to wreck the development of long-term solutions to energy (and related environmental) problems. In this regard, one should never say that nuclear energy is not for developed countries, in they are the ones hurt most my the vagaries of the short-term energy market as they strive to pull themselves up to even a small fraction of the energy usage and relate living conditions enjoyed by the industrialized countries. Government policy must regulate in a way that long-term energy options are developed and implemented in relative insulation from the vagaries of the market. Lastly (Alan Pasternak), while the market forces are short-term in nature, government actions must deal with institutional longevities and response times that can be influenced by shorter time scales relate to election cycles, through which the short-term nature of the market forces can be felt. Furthermore, environmental issues and related time scales can be viewed differently, as initiatives in one country can be opposed by other countries.

ROUND TABLE DISCUSSION

Moderator, Dr. Pierre Zaleski, Universite' Paris, Dauphine Place de Marechal de Lattre de Tassigny 75775 Paris Cedex 16 France

Annotator: Mrs. Behram N. Kursunoglu, Global Foundation, Inc. P. O. Box 249055 Coral Gables, Florida 33124-9055

Clinton Bastin (U.S. Department of Energy, Washington, DC); Gerald Clark (The Uranium Institute, London); Curt Mileikowsky (Switzerland); John Nuckolls (Lawrence Livermore National Laboratory, Livermore, CA); Ed Rodwell (Electric Power Research Institute, Palo Alto, CA); Richard L. Wagner, Jr., Los Alamos National Laboratory, Los Alamos, NM); and Bertram Wolfe (General Electric Corporation, San Jose, CA).

Each panel member participating in this round-table discussion was asked to address the following two questions: *a) Can the world sustain its economic and environmental needs without nuclear energy?; b) Can the use of nuclear energy be globalized with built-in security and safety measures?*

Generally, the consensus answers were "no" for question a) and "yes" for question b). A brief elaboration of responses from each of the panelists is given below, with the main differences amongst panelist being the relative seriousness of the actual proliferation problem as related to the civilian fuel cycle.

Burt Wolfe (General Electric Company, retired):
The answer to question a) is dependent on time scale, with no problem envisaged for the short run, particularly in the over-supply situation that characterized the present U.S. electrical-generation scene. Problems will occur later, as demand, resource, and environmental concerns loom larger than they are perceived at present.

Richard Wagner (Los Alamos National Laboratory):
The main concern with respect to question a) is how to encourage government support for a nuclear-energy option that is not crucially needed at present, but will have an increasing need in the future as global demand, resource competition, and environmental impacts grow. A means must be found to keep the nuclear-energy option alive, if not further development of the option, until it is seriously needed. More lucid arguments must be developed as to why the government role as a driver to keep open and to improve nuclear energy must be made and conveyed to both the public and the governments. Specific options for translating this need/concern into action were presented in the form of the following thought-catalytic questions: a) What is the role of the nuclear industry in making the required case?; b) What are the merits and demerits of allowing local nuclear

expertise to decay, with the plan/goal of importing the required "know-how" and technology at a later date from burgeoning overseas markets?, and c) What are the potential and feasibility of initiating an international program of R&D in nuclear energy?

Clinton Bastin
(U.S. Department of Energy): Either perceived or real environmental needs will increase pressure for increased use of nuclear energy. This increased use of nuclear energy will lead to the need for increased security in an environment of expanded use, while contributing to increased regional security relate to the satisfaction of local energy needs.

Gerald Clark (The Uranium Institute):
Displeasure with both questions was express because of the perceived negative and/or defensive tone that is telegraphed. Concern was express over the emphasis on proliferation from the civilian nuclear fuel cycle placed on the Los Alamos nuclear futures study, notwithstanding resonant concerns in other sectors that originally motivated the prominence of the proliferation issued in the Los Alamos study/assessment effort. While question b) was given a "yes" answer, it was noted that past studies at The Uranium Institute indicated that the actual cost of dealing with the proliferation issue adds 0.1-0.2% to the cost of reactor operations and 2-4% to the cost of reprocessing and MOX fabrication.

Kurt Mileikowsky (Switzerland, retired):
Global warming and nonproliferation are major elements in the world energy arena and will remain so for the next ~50 years. Reactor plutonium was viewed as having a strong potential for making nuclear weapons. Direct repository disposal of any form of plutonium requires elaborate international (IAEA) control and monitoring for 100,000s of years as the plutonium mining potential approaches a favorable level exponentially with time; to avoid the unrealistic possibility of monitored control in perpetuity, the need is great for efficient plutonium "incineration" and effective plutonium-inventory minimization.

John Nuckolls (Lawrence Livermore National Laboratory):
Any scenario without nuclear energy is wrong-headed on all accounts: a) the morality of burning oil and leaving a legacy of depleted petroleum and enhanced atmospheric carbon dioxide to future generations must be examined; b) future nightmare scenarios related to increased greenhouse-gas concentrations, dramatic changes in local and global climates, and political implications of increased terrorism may signal a wakeup call that only promotes an awaking into the midst of an ongoing nightmare. When these scenarios are viewed through both short-term and long-term lenses, time is on the side of nuclear energy (e.g., resource limits/depletion, greenhouse-gas accumulations, etc.). The present U.S. scene for nuclear energy is temporary, the short-term focus onto overseas markets is appropriate, and the driving down of nuclear costs for future domestic implementation is an important goal. Both cost and (public) attitudes towards health risks associated with radiation are decisive drivers for nuclear power, with recent/ongoing work on the human genome and the far-greater importance of biological/chemical elements in establishing health risks eventually leading to a return of rationality to that debate. A priority agenda must be formed on dealing with economic growth and the freeing of markets, along with the injection of increasing doses of pragmatism into crucial political decisions and debates (e.g., human rights may have to slip down the agenda in deference to increasing market interaction in/with China).

Edward Rodwell (Electric Power Research Institute):

Question a) received a strong "no" and question b) was given a strong "yes". In dealing with this answer for question a), significant increases in spent-fuel management will be required, with the greatest technological challenges occurring in reprocessing, and the fabrication and movement of MOX. The levels of security and safeguards will have to grow to meet these needs.

A Final Note from the Annotators

Robert Krakowski and José G. Martín

The focus of the International Conference on Technology for the Global Economic, Environmental Survival and Prosperity was to be on topics related directly related to environmentally attractive technologies for the long-term generation of electrical power. Almost by definition, this focus refers to so-called renewable energy sources, advanced fossil fuel plants, and nuclear power. The implicit goal was to establish a forum for communication, clarification of issues, and synthesis. The implicit hope was that "if a field was built, *they* would come."

It is clear from the above summaries that the professional interests and expertise of the conferees, and the nature of their presentations, gave the Conference a strong focus on the nuclear option, while recognizing the need for a broad and flexible spectrum of long-term energy options. Over and over again, the speakers themselves referred to the need to reach out not only to the public and the decision makers, but also to those whose ideas and perceptions of energy alternatives have been shaped by rigid interpretations of dangers and costs.

There is much agonizing in this country about the challenges our teachers face when trying to educate future generations faced with problems that are not of the teachers' making. In an analogous fashion, we can lament the challenges we all face when fora for rational discourse are ignored by those who could contribute to productive synthesis.

The teachers of America are to be thanked and complemented for not giving up their fight for our survival. In analogous fashion, Dr. Behram Kursunoglu and the Global Foundation, Inc., are to be thanked and complemented for keeping the forum open, and the welcome mat extended. Their efforts help preserve the legacy of rationality and scientific positivism that made dissent viable for all of us. Their efforts also help maintain the hope for scientific progress and human betterment; it is that hope that makes it all worthwhile.

CHAPTER I

GLOBAL ENERGY DEMAND AND ENERGY SECURITY PROJECTIONS IN THE COMING CENTURY: GLOBAL ECONOMY, ENVIRONMENT, RESOURCES, AND POPULATION ISSUES

THE REALITIES OF SUSTAINABLE DEVELOPMENT

IN THE TWENTY-FIRST CENTURY

William F. Martin

Chairman
Washington Policy and Analysis, Inc.
1025 Thomas Jefferson Street, NW
Washington, DC 20007

INTRODUCTION

Good morning. It is a pleasure to be here today. I would first like to thank Ed Arthur of the Los Alamos National Laboratory for introducing me to the Global Foundation and to Dr. Behram Kursunoglu for inviting me to speak and participate in this weekend's conference.

This is the first opportunity for me to participate in a Global Foundation conference, and from what I have read and learned about the organization, your concerns are global in scope, and stress the challenges we face in meeting long-term energy demand—with a special emphasis on nuclear power.

I share your beliefs as to the necessity and importance of addressing long-term energy futures. About a year ago, Paul Volcker asked me to take the lead in preparing a study on energy security for the Trilateral Commission—a group of distinguished individuals from Europe, North America, and Japan.

My co-authors were Helga Steeg, former Executive Director of the International Energy Agency (IEA) and Ambassador Ryukichi Imai, whom I am sure many of you know as one of Japan's leading authorities on non-proliferation and formerly Japan's Ambassador to the Conference on Disarmament in Geneva, to Kuwait, and to Mexico.

Technology for Global Economic and Environmental Survival and Prosperity
Edited by Kursunoglu *et al.*, Plenum Press, New York, 1997

Our report, *Maintaining Energy Security in a Global Context,* which was published in September, addresses the energy and environmental challenges we face over the next 15 years. This morning I would like to use the Trilateral Commission study as a launching platform for examining our long-term energy supply and demand future through 2050.

I have titled my speech: "The Realities of Sustainable Development in the Twenty-First Century." Put simply, what type of energy system needs to evolve if we are to achieve sustainable development in the next century? Furthermore, what role should nuclear power play in helping us achieve our goal of a sustainable future?

The term "sustainable development" has become one of the most widely used environmental and political concepts over the past several years. It first appeared in 1987 when the World Commission on Environment and Development (commonly referred to as the Bruntland Commission) released its report, *Our Common Future.* [1] Sustainable development was defined as: "meeting the needs of the present without compromising the ability of future generations to meet their own needs."

It is a concept that gained momentum in the late 1980s and played an influential role in bringing together more than 150 countries to sign the Framework Convention on Climate Change in Rio in May 1992—committing themselves to achieving a stabilization of greenhouse gas concentrations in the atmosphere at a level that would minimize human interference with the global climate system.

The Clinton Administration, led by Vice President Gore, has made sustainable development a core element of its environmental policy. In 1993 a presidential commission was established—the President's Council on Sustainable Development—to study and improve government and private sector activities that would ensure sustainable development.

The panel's final report, *Sustainable America,* which was released in February 1996, included a number of recommendations, including a commitment that "our nation will protect its environment, its natural resource base, and the functions and viability of natural systems on which all life depends.[2] I congratulate the Administration on this fine piece of work.

This morning I would like to dig a little deeper into this concept, expand its definition—to include the element of national security—which I believe plays a vital role in achieving any type of sustainable future, and look at what the realities are in achieving a sustainable future worldwide.

In addressing the realities of sustainable development in the twenty-first century, I have divided my remarks into five areas:

- First: What is the nature of the national security threat—which I believe is an important part of the sustainable development concept—over the next 15 years, and to what extent will it involve the Persian Gulf?

---

[1] World Commission on Environment and Development. *Our Common Future* (Oxford: Oxford University Press, 1987).
[2] President's Council on Sustainable Development, *Sustainable America* (Washington, DC: U.S. Government Printing Office, 1996).

- Second: What are the likely patterns of energy supply and demand over the next 15 years? Do they suggest we are heading for a sustainable future?
- Third: What are the likely patterns of energy supply and demand over the next 50 years? Are they sustainable? If not, what needs to be done?
- Fourth: What specific contributions can nuclear power make in helping us achieve sustainable development? If nuclear power is to play a vital role in the long-term, what challenges does it need to overcome in the short-term?
- Fifth: I would like to conclude my remarks by summarizing my main points and emphasizing what type of key messages concerning nuclear power should be conveyed by conferences like this to senior policymakers in the United States, other OECD countries, and the world as a whole.

PART I: NATIONAL SECURITY AND SUSTAINABLE DEVELOPMENT—THE CASE OF THE PERSIAN GULF

The National Security Threat of Rising Dependence on Persian Gulf Oil

Although the concept of sustainable development is one that is closely tied to the effects of our economic development on the environment and how this will affect future generations, I think it can be broadened to include national security. A well defined and structured national security policy—of which energy is an important sub-component—is a vital part of our planning for future generations. I believe we face a daunting energy-related national security challenge within the next 15 years. That challenge is our increasing reliance on oil from the volatile Persian Gulf region.

Many in the energy community discount the nature of the energy security threat that is before us. They argue that improvements in oil recovery technology, stable fiscal regimes, and the emergence of a global oil market lessen the significance of Persian Gulf oil producers and thereby minimize the concept of energy security. Furthermore, persons claiming this view believe that in the event of another crisis in the Persian Gulf, such as an attack on Saudi Arabia or Kuwait, the U.S. would respond immediately by sending 500,000 troops to the region and restore order, thus ensuring the stable supply of oil to world markets. I disagree wholeheartedly with this point of view. In fact, I believe we are in an even more perilous situation than that we even faced during the 1970s, 1980s, and early 1990s—a period during which we experienced an oil disruption due to geo-political events on average every 5 years. There are several reasons as to why I believe our increasing reliance on oil from the Persian Gulf poses a significant national security threat:
- Rising Oil Demand: Oil demand is rising at an unprecedented rate and will continue to be the world's most important energy source. It will account for over 40 percent of global primary energy supply in 2010. Without stable and affordable supplies of oil, the global economy will grind to a halt. Much of this oil demand over the next 15

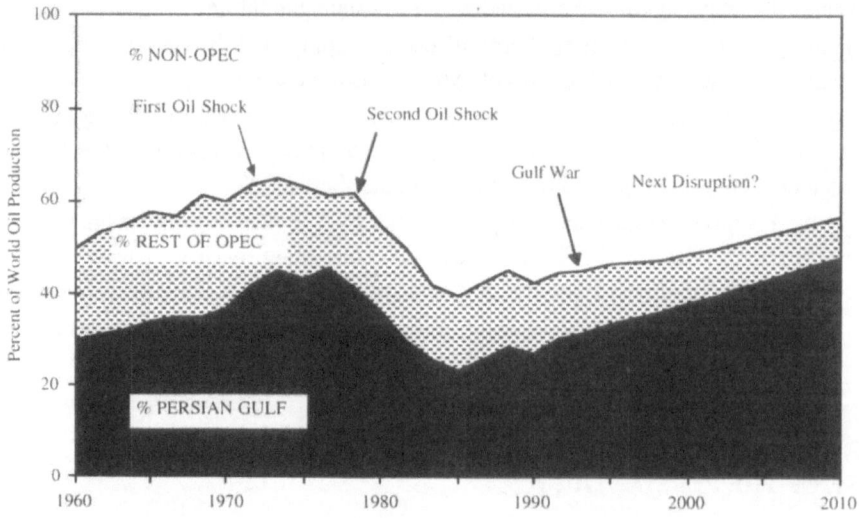

Source: GEMS Global Energy Supply and Demand Model.

Figure 1. World Dependence on Persian Gulf Oil

years will originate in the rapidly industrializing economies of East Asia. They will account for a greater increase in annual oil demand than the whole of the OECD. This growth in oil demand is being fueled largely by an expanding transportation sector.[3]

- Non-OPEC Production: OECD regions, especially the North Sea and parts of North America, have experienced impressive gains in oil recovery rates in the more mature fields. Other parts of the world (non-OECD) are also increasing oil production as the result of more stable fiscal regimes and technological improvements. Nevertheless, with projected increases in worldwide oil consumption, steady non-OPEC oil production represents a declining share of total world oil supplies.

- Persian Gulf Swing Producers: Despite the positive prospects for steady (instead of declining) non-OPEC supply of oil, the world's incremental oil demand will have to be met by OPEC producers—and within OPEC by the Persian Gulf producers: Iran, Iraq, Kuwait, Qatar, Saudi Arabia, and the United Arab Emirates. As demonstrated in Figure 1, it is the percentage of world oil supplies provided by the Persian Gulf producers (not OPEC overall) which we project rising to mid-1970s levels around 2010.

Obstacles to Stable Oil Supplies from the Persian Gulf Region

Certainly the Persian Gulf swing producers—Saudi Arabia, Kuwait, Iran and Iraq—could produce this amount of oil based on their reserve potential, but there are serious obstacles:

---

[3] As a result of high economic growth, a rapid long-term expansion in the global transportation sector is occurring. Given current low vehicle densities in non-OECD countries, there is an enormous potential for expansion in the world's transportation sector. Average growth in the transport sector is expected to be 1.6 percent in OECD countries, 2.4 percent in the former Soviet Union, and 4.5 percent in the rest of the world (ROW). The highest growth is projected for East Asia, at more than 6 percent annually.

- Unstable Investment Climate: Governments in the region face daunting socio-economic challenges in the coming decades. Past economic policies, based on state-directed investments and subsidies paid by oil rents, are no longer viable. The challenges of restoring economic growth, restraining population growth, creating jobs, providing food, conserving water, and protecting the environment will require large sums of money. Only the private sector, both domestic and foreign, can provide the necessary funds. Yet if such investments are to occur, stable and predictable "rules of the game" for private investors must be established. Private sector investment will be a crucial component of any expansion in the oil production capacity of the Gulf states.

- External Threats: Defending against *external* threats to a key producing country—such as Iraq's invasion and annexation of Kuwait in 1990 or the war between Iran and Iraq in the 1980s—is of central concern and is becoming more challenging. Advancements in military technology, especially the development of weapons of mass destruction (nuclear, biological, and chemical) by pariah states, raises questions about future U.S. military activity in the region with the purpose of defending against external threats.[4] The U.S. Department of Defense has recently released a report outlining Iran's commitment to developing weapons of mass destruction and the threat this imposes on the region. As Map 2 illustrates, present day Iranian ballistic missile technology would enable it to strike targets in neighboring countries, including oil installations and ports in Saudi Arabia.[5]

- Internal Threats: Defending against *internal* threats—such as the collapse of the Shah's regime in Iran, which brought about the second oil shock in the late 1970s—is one that is difficult for outside nations, such as the United States, to influence. Several monarchies in the Gulf, such as Saudi Arabia, a re in the midst of transferring power to a younger generation. If we support the current monarchical structures, which are often undemocratic, are we supporting long-term revolutions, as was the case in 1979 in Iran? On the other hand, if we withdraw our support for these established regimes, would they fall into the hands of radical clerics, as is the case of Iran today?[6]

Optimistic Viewpoint

What could change this picture, remembering that our outlook period extends to 2010, which is not a long time in terms of energy planning?

---

[4] Meeting National Intelligence Council (NIC) on Future Vision 2010/Middle East Seminar. October 1, 1996.

[5] Office of the Secretary of Defense, *Proliferation: Threat and Response*, (U.S. Department of Defense, April 1996).

[6] Some have suggested that despite the chaos and confusion from an internal revolution, the successor regime would still have to sell oil in order "to eat." This may be the case at the end of the revolution, but what happens during the interim fighting that may involve rival factions and last several months? A sudden drop in oil production from any of the major Persian Gulf producers for even a period as short of 3 months would wreak havoc on the global economy. Recall that the oil crises of the 1970s became the international debt crisis of the 1980s.

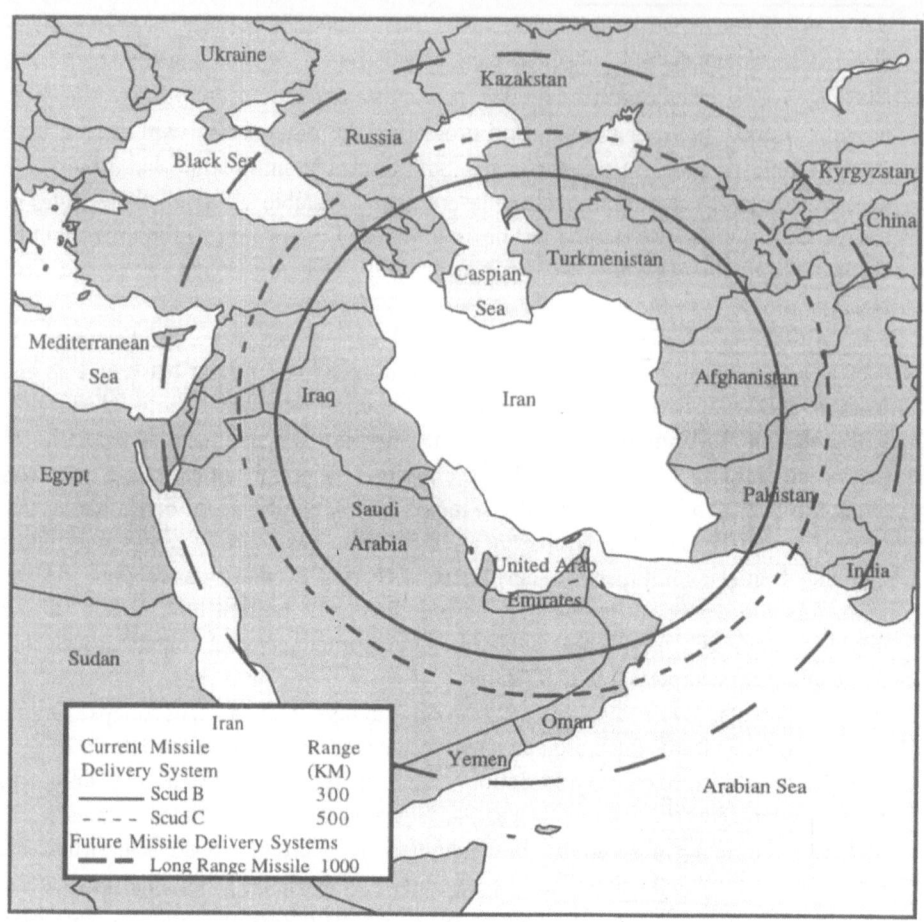

Map 1. Iranian Ballistic Missile Technology

- Lower Oil Demand: Oil demand could be lower than we project. However, our GEMS Model projections are based on conservative estimates of future economic growth. These projections are actually lower than almost all other comparable models. For example, we estimate Chinese economic growth at a relatively low average rate of 6 percent during the outlook period, compared with OECD's estimate of 8 percent and the World Bank estimate of 10 percent. We also assume improvements in energy efficiency. Therefore, if anything, we have erred on the low side of energy [oil] demand. It could, in fact, be higher.
- Technological Improvements: Breakthroughs in oil recovery technology are an important reason for the better-than-predicted non-OPEC oil production over the past decade. Improvements in recovery rates have brought about reductions in cost to the point where North Sea recovery costs are less than half what they were a decade

ago. In our model, however, we account for improvement in recovery, and in fact, we show an increase in non-OPEC production for the forecasted period (1990-2010). Many forecasters will say that this is too optimistic, but we believe it is realistic given advancements in 3-D seismic, horizontal drilling and other technologies. Nevertheless, the rate of world demand for oil easily outstrips the capability of non-OPEC countries to meet that demand, resulting in rising dependence on Persian Gulf producers.

- Alternatively Powered Vehicles: The transportation sector is almost totally dependent on oil. Demand for oil will rise corresponding to the rapid growth projected in the transportation sector. One way of slowing that growth in oil demand would be the wider use of alternatively powered vehicles. In the short-term, this could involve greater use of natural gas, and other fossil fuel derivatives. Over the long-term, electrically powered vehicles could significantly reduce our oil consumption; however, unless a technological breakthrough occurs in battery technology (which would both extend their range and reduce their cost), electric vehicles will not make any significant impact on the transportation sector until well after 2010. Furthermore, it takes time to build distribution infrastructure and the rate of auto replacement on average is only once every 10 years. Therefore, it may be 20 to 30 years before we see a significant (say 25 percent) increase in market share of alternatively powered vehicles.

- Emergence of the Former Soviet Union as Major Oil Exporter: During the 1980s, Russia was one of the largest crude oil producers in the world. In 1987-88, its production peaked at 12.5 million barrels per day. Today, its production level is half that amount at around 6 million barrels per day. The combination of low domestic prices, non-payments by consumers, high taxation of producers, and an uncertain fiscal and legal regime, has caused a collapse in drilling activity and new field development. Oil production has probably bottomed out and will slowly begin to rise, but it will not reach its high production levels of the 1980s until well into the next century. Another potential supplier of oil to world markets in the area of the old Soviet Union is the Central Asian and Caspian Sea region. Oil reserves in this area represent a major new source of oil for the 21st century. Before this becomes a reality, however, the oil needs to be transported to markets in Western Europe and Asia through a volatile belt of countries and districts. There is much uncertainty surrounding the development of adequate pipeline routes to consumer markets.[7]

- Economic and Political Stability in the Middle East: Economic and political stability in the Persian Gulf region is an essential element of maintaining energy security. Each of the oil market disruptions of the 1970s, 1980s, and 1990s was linked to instability in the Gulf region. I certainly hope peace and economic stability reign throughout the region, but we should not count on it.

---

[7] If pipelines are built according to plan, it is projected that oil exports from the Caspian Sea could be as high as 700,000 barrels per day beginning in 1997.

Bottom-Line Assessment: the Path Toward a Sustainable Future?

The world is steadily heading toward greater reliance on the Persian Gulf for its principal energy resource—oil. Indeed almost half of the world's oil may have to come from this volatile region within the next 15 years. The region is poised to undergo significant change over the next decade. The aging leaders, the increasingly apparent dissatisfaction on the part of youthful and frustrated populations, and the growing assertiveness of movements and ideologies that contest the legitimacy of aging monarchies and civilian dictatorships all suggest that the region is ripe for major political upheavals. Against this backdrop, prudent policymakers should be asking themselves several questions:

- Are we maintaining a well-balanced energy portfolio—one that will help insulate us from the effects of another oil shock?
- What political/military options do we have in responding to a crisis in the Persian Gulf?
- What would be the costs, in terms of economics and national security, of another major oil shock?
- Is our present national security policy sustainable?

Some people may say that my perspectives on rising dependence on the Persian Gulf are overly pessimistic. However, I like to point out that in the field of national security, we never plan for good news.

PART II: WORLD ENERGY SUPPLY AND DEMAND OVER THE NEXT 15 YEARS

Explanation of GEMS Model

Having addressed the national security challenge associated with rising dependence on oil from the Persian Gulf, I would now like to examine the likely patterns of energy supply and demand over the next 15 years and determine if these patterns are leading us toward a sustainable future.

Our tool for examining energy demand and supply over the next 15 years is the GEMS Model. The GEMS approach was first developed at MIT by Carroll Wilson and the Workshop on Alternative Energy Strategies, where I served as a program officer 20 years ago. Jean Couture of France, who I believe is here today, contributed significant time and effort to this project. It is a relatively "assumption driven" model, based on a number of factors regarding economic growth, oil price, and fuel preferences. It includes estimates of energy supply and demand for 11 countries, 10 regions and a global aggregate. It is therefore possible to look at fuel mix shifts and to estimate what it would take to reduce dependence on a particular fuel for security or environmental reasons. The model has over one million cells in an Excel spreadsheet format. We start with economic growth and here we include private estimates of the OECD secretariat. These estimates tend to be lower than member government estimates. We have conservative estimates for world economic growth over the next 15 years—2.3 percent for OECD countries and 5.3 percent for non-OECD countries. Estimates of economic growth for the rapidly industrializing countries are somewhat higher than that of the more "mature" OECD economies. This is due, in part, to free trade which has encouraged a shift in the world's manufacturing capability from OECD to the rapidly industrializing nations. Our GEMS Model looks at this from an energy point of view and evaluates energy demand, for instance, as the steel industry shifts from the U.S. to Japan to Korea to Thailand to India. Free trade has brought prosperity to industrializing countries and cheaper goods to OECD countries, where service economies actually help reduce our energy demand.

Global Energy Supply and Demand (1990-2010)

Our GEMS Model projects that world energy demand will likely increase by approximately 40 percent between now and 2010. This increase is being fueled by several factors: economic growth, population growth and urbanization, greater electrification, and expansion of personal mobility. As shown in Figure 2, OECD countries will constitute a smaller share of much greater world energy consumption than was the case in the early 1970s. Energy systems do not change much in 15 years, and accordingly, as shown in Figure 3, the GEMS Model projects that the world will continue to rely on fossil fuels to meet the majority of its energy requirements.

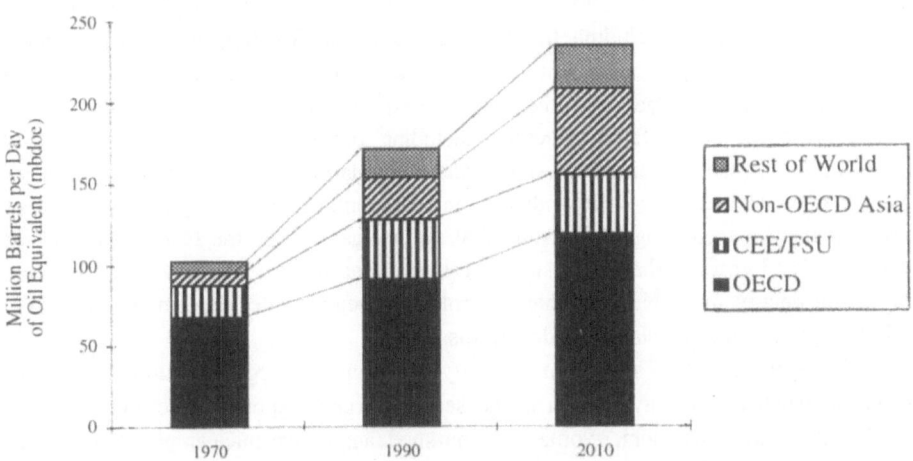

Source: GEMS Global Energy Supply and Demand Model.

Figure 2. Energy Consumption by Region (1970-2010)

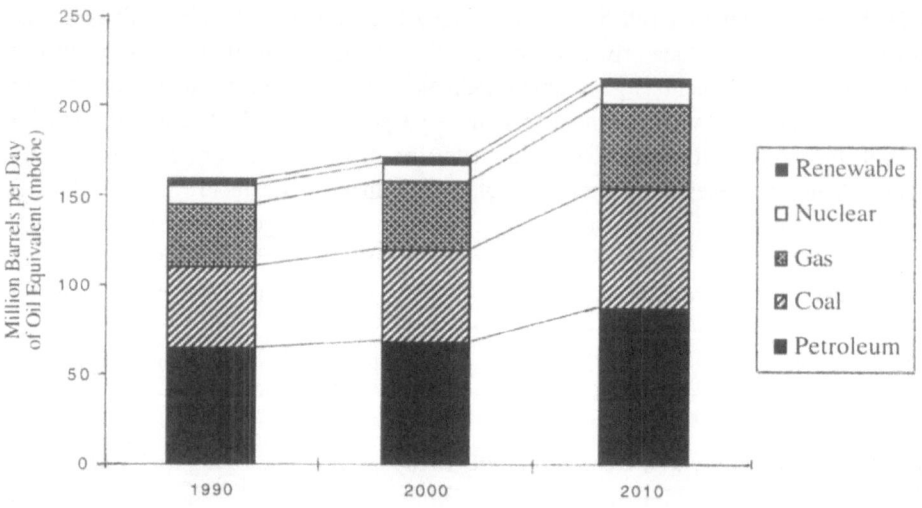

Source: GEMS Global Energy Supply and Demand Model.

Figure 3. World Primary Energy Demand by Fuel (1990-2010)

Regional Energy Supply and Demand (1990-2010)

Now that we have examined energy demand and supply patterns at a global level, I think it will be helpful to examine the regional energy demand and supply structure.

North America (including the U.S., Canada, and Mexico) has a large natural resource endowment and consequently, many different energy options. Natural gas has been the fastest growing energy source over the past several years. This has come about as a result of several different factors, including the implementation of the North American Free Trade Agreement (NAFTA) and deregulation of energy markets in general. Changes in the U.S. electric utility industry, environmental regulations and improvements in combined cycle technology will ensure that natural gas will be the fastest growing fuel over the next 15 years. Other fuels, such as oil, coal, and nuclear will retain their share in the overall energy mix. Nuclear energy's role will begin to decrease in the post-2010 period as many plants are slated for decommissioning.

European countries represent a variety of different energy situations. Some are producers (such as Norway, which is the second largest exporter of crude oil in the world); others are consumers. Some have pursued aggressive nuclear power programs, such as France and Belgium, and others such as Italy have relied on oil for electric generation. We expect the energy demand/supply structure to remain largely reliant on fossil fuels over the next 15 years, with natural gas as the fastest growing fuel source.

Over 80 percent of Japan's energy needs are met by imports—including all of its oil, about 70 percent of which originates in the Middle East. Consequently, diversifying its energy supply structure is the primary objective of Japan's energy policy. Expanding its nuclear energy program is a principal part of this plan, including the development of a closed fuel cycle. The Japanese government's target of having nuclear energy provide approximately 40 percent of electric generation by 2010 will require the addition of approximately 15 nuclear reactors.

The rapidly industrializing countries of Asia[8] will account for a large percentage of the increase in global energy demand over the next 15 years. Sustaining real economic growth rates of between 6 and 8 percent will require massive amounts of energy. We expect these countries to rely on the most inexpensive fuels—largely oil and coal—to satisfy their expanding energy appetites. For example, we project that China will rely on coal for 75 percent of its primary energy share in 2010.[9]

This raises an important, yet sensitive issue: the need for the industrialized [OECD] countries to take the lead in using the most advanced energy technologies, which are also the most expensive. Less industrialized countries will rely on the cheaper energy technologies, just as we did during our earlier stage of development. It is important to note, however, that these "cheaper energy technologies" will still be much more efficient than the technologies we used 50 years ago. A coal plant being built in China today is far more advanced than one that was built in the Ohio Valley just after World War II.

Now that we have examined energy and demand patterns over the next 15 years, what trends emerge?

The energy portfolios in OECD countries will remain relatively balanced, with a mix of fossil fuels, nuclear, and renewables. However, beginning in 2010, nuclear generating capacity in many of these countries, with the exception of France and Japan, will decline—increasing our reliance on fossil fuels.

The rapidly developing countries, especially those in Asia, will use an increasing percentage of fossil fuels, especially coal, because of its abundance and low cost. By 2010, fossil fuels will account for 80 percent of East Asia's primary energy supply.

Bottom-Line Assessment: the Path Toward a Sustainable Future?

Our analysis of energy supply and demand patterns over the next 15 years points toward some troubling conclusions. The world, undergoing a profound economic expansion, is turning toward the most inexpensive energy sources available. This trend is not putting us on the right path toward a sustainable future, which so many governments, including our own, have made such a top priority. Two of the most troubling trends are in the areas of national security and environmental quality—principal elements of any strategy for achieving sustainable development.

I have already alluded to what I think are the national security consequences of our continued reliance on oil from the volatile Persian Gulf. I believe we face the prospect of another oil market disruption within the next 3 to 5 years. Recall that the shortfall

---

[8] Asia is defined as: China, Hong Kong, India, Indonesia, Malaysia, Philippines, Singapore, Republic of Korea, Taiwan, Thailand, and Vietnam.

[9] China will be the single largest source of greenhouse gas emissions during the next 15 years.

brought forth as a result of the Iranian revolution in 1979 involved only 5 percent of world oil supplies and lasted only three months. Yet it brought forth profound economic dislocation, both in industrialized countries, which plunged into deep recessions, and also for developing countries, which suffered severe balance of payments problems.

Expanded energy, both in the production and consumption phases, brings forth a host of environmental challenges—ones directly related to the theme of sustainable development. The two that I will touch upon briefly—acid deposition and global climate change—are the direct result of our use of fossil fuels.

Acid deposition, which first became a major concern in the highly industrialized regions of Europe and North America during the 1970s and 1980s, is beginning to become evident in many areas of East Asia—especially along the southern coast of China, the Korean peninsula, and Japan. Available monitoring shows that the acidity of rainfall has been rising dramatically in some areas of the region.

If counter measures are not taken, acid deposition in many areas will increase by more than a factor of five and exceed the levels observed in the most polluted areas in Central and Eastern Europe.[10] Table 1 shows the current and projected emissions of sulfur dioxide for Europe, North America and Asia. The total projected sulfur dioxide emissions for Asian countries in 2000 and 2010 far exceed North America and Europe combined.

Table 1. Current and Projected Sulfur Dioxide Emissions by Region (million metric tons)

Region	1990	2000	2010
Europe	38.0	22.0	14.0
North America	21.0	15.0	14.0
Asia	34.0	53.0	78.0
China	22.0	34.0	48.0
India	4.5	6.6	10.9
Other	7.5	12.4	19.1

Source: RAINS-ASIA Program.

Although uncertainty surrounds the theory of global climate change (regarding the extent and speed of such change, its overall effects and regional distribution, and the cost and effectiveness of efforts to prevent, slow down, or adapt to change), the magnitude of the risk led more than 150 countries to sign the Framework Convention on Climate Change (the Rio Convention) in 1992.

Despite the fact that political leaders in OECD countries have made commitments to stabilize their greenhouse gas emissions (at 1990 levels) by 2010, these commitments

---

[10] Wes Foel et al, *RAINS-ASIA: An Assessment Model for Air Pollution in Asia*, Report on the World Bank Sponsored Project: "Acid Rain and Emission Reductions in Asia" (Washington, DC: World Bank, 1995). Some experts have noted that acid deposition could impact China's agricultural productivity, threatening to turn China into a net importer of rice in the next century.

will not be achieved. The necessary energy policies to meet these targets are not being implemented. In fact, our GEMS Model/IEA projections demonstrated in Figure 4 illustrate that energy-related carbon dioxide emissions in 2010 will be over 30 percent higher than the 1990 level.

If the issue of global climate change is to be taken seriously over the long-term, a fundamental change will have to be made in the global energy structure, with increased reliance on non-fossil fuels such as nuclear and renewable energy sources. In this section I have tried to outline what the likely patterns of energy supply and demand are over the next 15 years—a relatively accurate analysis given that 15 years is not a large time horizon for energy systems. We have seen that our path to the twenty-first century is one that is becoming increasingly reliant on fossil fuels. This reliance on fossil fuels raises significant energy security and environmental challenges—ones that would seem to be in direct conflict with our goal of sustainable development.

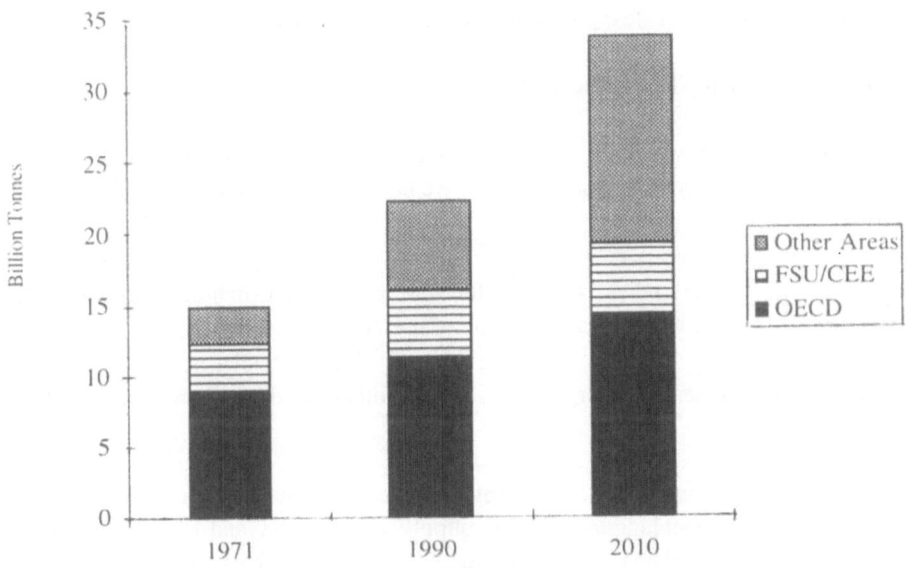

Source: GEMS Global Energy Supply and Demand Model and IEA.

Figure 4. World Carbon Dioxide Emissions

PART III: ENERGY SCENARIOS—FIFTY YEARS AHEAD

In October I participated in a seminar organized by Dr. Jack Gibbons, the President's Chief Science and Technology Advisor, and Director of the White House Office of Science and Technology Policy. Jack was kind enough to invite me to the White House to provide him and his senior staff with an overview of our Trilateral Commission report.[11] Dr. Gibbons made several interesting comments at our meeting, including his

---

[11] Meeting with Dr. Jack Gibbons, Science Advisor to the President and Director of the White House Office of Science and Technology Policy (OSTP). October 22, 1996.

assessment how we should examine our long-term energy future—the period 2010 to 2050. Jack pointed out that in examining long-term energy supply and demand, we first need to admit that there will be no "silver bullet" that will solve all our problems. Our energy solution, he noted, will involve several different technologies, including greater penetration of renewables,[12] more energy efficiency and better conservation, expanded use of clean-burning natural gas, and advanced nuclear energy technologies. He said that if we want to see what type of role nuclear power will play, let us first try to determine the contribution of other energy sources and technologies. How much energy will renewables and natural gas be able to provide? Will greater energy efficiency and more conservation lead to reduced total energy demand? What type of transportation sector will be in place—fossil fuel-based or reliant on alternative energy sources (electrical powered vehicles)? Then, he stated, we will have a better idea of the need for nuclear energy— what amount of nuclear energy will be need to fill the gap? Dr. Gibbons' observations are important and we were challenged to estimate the fuels requirements of a sustainable development future, including the range required of "clean and sustainable fuels" such as natural gas, renewables, nuclear power and greater energy efficiency.

Factors Driving Long-Term Energy Demand

This morning I would like build on Dr. Gibbon's remarks and present to you two different long-term energy scenarios that have we have generated from the GEMS Model. The two cases are:

- Business as Usual: relies on fossil fuels to provide the majority of our energy resources. In other words, a continuation of present policies; and
- Sustainable Growth: enhances the role of renewables, natural gas, and nuclear, and includes improvements in energy efficiency. In other words, a digression from fossil fuels.

In assessing long-term energy supply and demand futures, there are a number of factors which could affect long-term energy supply and demand, and I have tried to take these into account in building our projections.[13] In making our estimates for economic growth, we realize that there will be more rapid economic growth in some periods and much lower growth during others. Consequently, our economic growth for the 50 year period is 2.4 percent. This is about the average economic growth of the last 50 years. Population growth will be another influential factor—especially the increasing urbanization of the world population.

Our estimate for population growth over the next 50 years is approximately 1.7 percent, which is considerably lower than the average growth rate of 2.1 percent from

---

[12] Included as renewable energy technologies are: hyrdo, solid biomass, photovoltaic cells, wind power, geothermal, passive solar, and hydrogen.

[13] We use the same estimates of economic and population growth in both scenario cases.

1971 to 1992. As demonstrated in Figure 5, even with low growth rate expectations, the world's population growth could almost double in the next 50 years.[14]

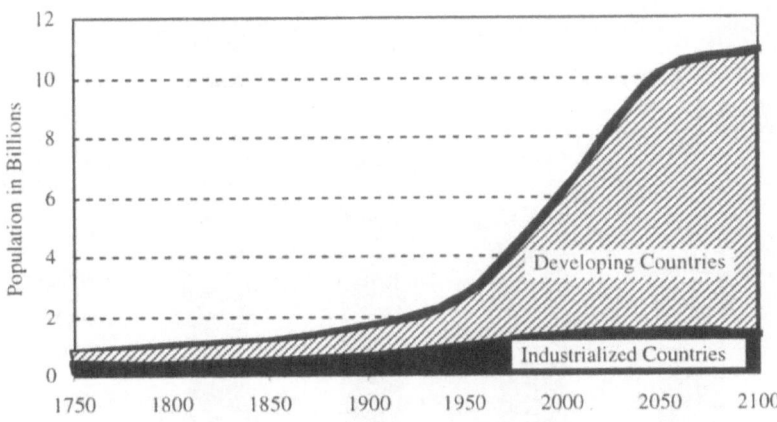

Source: World Bank.

Figure 5. World Population Growth (1750-2100)

Business as Usual

First, let me discuss the business as usual strategy, which is based on a continuation of present policies and market conditions.
Our assumptions for business as usual are:

- Continued reliance on oil in the transportation sector.
- Continued reliance on coal for electrical generation, especially in countries like China and India.
- Continued progress in energy efficiency in the transport, industry, and residential sectors.
- Continued growth in use of natural gas, including development of long distance pipelines.
- Continued increase in use of renewables, but no serious market penetration.
- Continued decline in nuclear energy generating capacity in OECD Europe and North America, but expansion of generating capacity in Asia.

---

[14] By 2025, the world population is projected to total 8.3 billion people, or about 45 percent more than the current estimated population of 5.7 billion. By 2050, world population projections reach 10 billion.

With what type of energy future does this present us? As demonstrated by Figure 6, it is one in which we continue to use enormous amounts of fossil fuels—especially oil and coal. I don't think this type of future is sustainable, especially from a national security and environmental quality perspective.

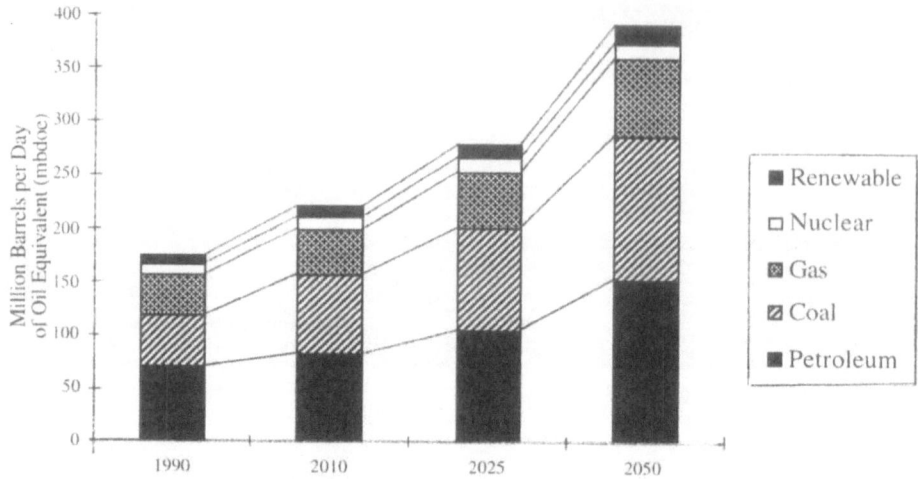

Source: GEMS Global Energy Supply and Demand Model.

Figure 6. World Primary Energy Demand by Fuel (1990-2050) --Business As Usual

Sustainable Growth

That is why we take a hard look at our second scenario—sustainable growth—and ask, what must be done today to achieve a sustainable future? Given our huge energy supply and demand matrix, we try to force the model to give us results which are sustainable. Our principal assumptions for sustainable growth are:

- Improvements in energy efficiency and conservation.
- Growth in role of renewables in electricity generation sector.
- Augmentation of natural gas in transport, industrial, commercial, and electrical generation sectors.
- Expansion in nuclear generating capacity—more than threefold by 2050.
- Decline in use of oil and coal (as a percentage of total energy demand) in all sectors. Oil and coal production does not exceed 20 percent of production rates in 2010.
- Increase in use of electric vehicles in transportation sector (30 percent market share by 2050).

Figure 7 illustrates our projections for the sustainable growth scenario.

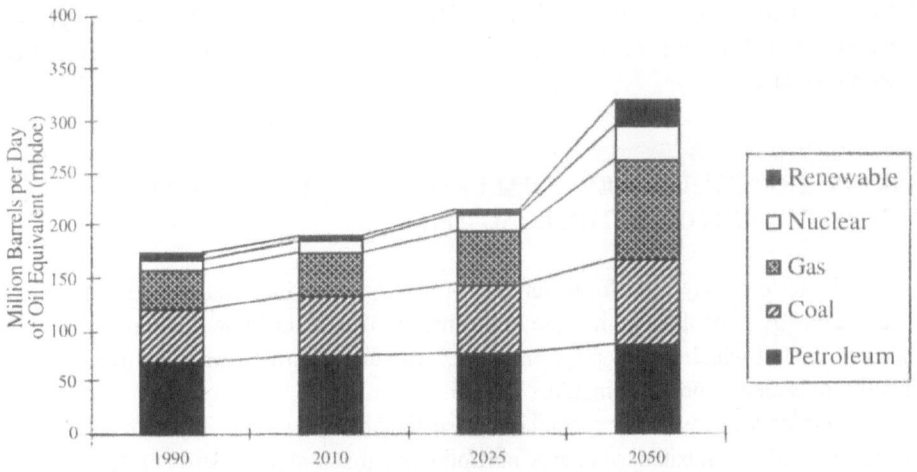

Source: GEMS Global Energy Supply and Demand Model.

Figure 7. World Primary Energy Demand by Fuel (1990-2050) -- Sustainable Growth

What does this energy projection tell us about our future? We can see from the sustainable growth scenario projection that fossil fuels still form a significant part of our energy portfolio, but far less than was apparent in the case of business as usual. Clean burning natural gas becomes the most important fossil fuel. We also see that renewables assume an important part in our energy supply picture.

What about nuclear power? Even with nuclear power only slightly increasing its percentage share of total electricity generated (from approximately 16 percent in 1990 to 20 percent in 2050), we see that nuclear energy generating capacity will have to increase about three-fold—from 325 GWe in 1990 to over 958 GWe in 2050.

This conclusion will not come as a surprise to this audience. What often surprises me is why environmental groups, so concerned about global climate change and acid deposition, do not embrace the nuclear energy option, a key element in any strategy designed to achieve sustainable development. Our sustainable growth scenario does not assume that nuclear energy will be the silver bullet that will overcome the energy challenges of the future. We are only assuming its role is slightly increased; yet, under this scenario, there would have to be an almost three-fold increase in nuclear generating capacity.

I think our snapshot of possible long-term energy supply and demand scenarios tells us that we have to maintain a well balanced portfolio of energy sources. It will be important to continue research and development on promising renewable technologies, especially biomass and photovoltaics. Natural gas we see as the fast growing energy source, largely because of its clean-burning attributes and availability. But this will require huge investments in natural gas infrastructure projects to get the gas to markets. We also underline the necessity of improving energy efficiency and promoting more conservation. Finally, we see nuclear power continuing to play a vital role in the electrical generation sector.

PART IV: NUCLEAR POWER: CHALLENGES IN THE SHORT-TERM AND OPPORTUNITIES IN THE LONG-TERM

We have clearly seen from our last projections the importance of maintaining a balanced energy portfolio. In the case of business as usual, our future is dominated by oil and coal, both of which raise serious energy security and environmental quality problems. This future is clearly not sustainable.

This leaves us with our second case—sustainable growth. If we are to achieve this future, we need a well balanced energy portfolio, one that includes a strong role for natural gas, renewables, and nuclear.

Short-Term Challenges

If nuclear power is to play a part in our sustainable growth scenario—which projects a threefold increase in generating capacity—there are several short-term challenges that it must overcome. These are:

- Finding a solution to the problem of long-term storage of waste.
- Ensuring that nuclear energy is economically viable.
- Maintaining a high safety record, particularly in the former Soviet Union and rapidly industrializing countries.
- Meeting non-proliferation objectives.

The single largest challenge facing the nuclear power industry in the United States, and I believe the world, is the problem of long-term storage of spent nuclear fuel and other high-level radioactive wastes from commercial nuclear power installations. If this challenge is not overcome, the future of nuclear energy in this country will be in peril. We must continue our efforts and funding for the completion and siting of the permanent waste depository at Yucca Mountain.

Internationally, we should seek not only individual nation sites, but also consider the concept of an international monitored retrievable storage site (IMRSS), which would enable us to share expertise and expenses; and meet economic, non-proliferation, environmental and safety concerns.

The economics of nuclear energy, compared with other energy choices, will determine whether or not the United States will replace the existing generation of nuclear power plants. In order to improve the economics of nuclear energy, we need to improve the regulatory environment. In Japan, it takes about 5 to 7 years to complete a plant. In this country, it can often take more than 10 years.

The Nuclear Regulatory Commission has made important steps in streamlining licensing procedures and expediting the licensing process for pre-approved standard designs. These efforts at streamlining our burdensome regulatory environment must continue. Many still believe that even with regulatory reforms, the cost of nuclear energy is still much too high. However, I like to view the economics of nuclear energy from a national security perspective—one that factors in the externalities of fossil fuels.

For example, although oil may appear inexpensive at $20 to $25 per barrel, there are many costs which are not factored into the price, such as the cost of maintaining a military force ready to intervene at a moments notice in the Persian Gulf.

Coal too is cheap. But what about its social costs—specifically, its impact on the environment? What if we were to include in the pricing structure for fossil fuels the potential ramifications of global climate change?

Nuclear energy is the only energy form in which all factors are inputted into its cost—from enrichment of natural uranium to the disposal of waste and decommissioning of nuclear plants.[15]

I like to point out that electric utilities in Japan are criticized for their expensive electricity. Global competitors such as Toyota and Fujitsu might argue that the high cost of electricity in Japan, brought about in part due to new construction of nuclear power plants, may hinder their economic competitiveness. And looking only at the electricity sector, they may be right. But think of Ford and IBM. They may pay less for electricity, but they, indeed all citizens and companies, pay a tremendous premium to maintain a $300 billion per year defense budget. We all pay for energy security. It just comes out of different pockets. Let me say, however, that Toyota does benefit by U.S. troops stationed in the Gulf. I would also argue that Ford and IBM will benefit in the long-run by improvements in the Japanese nuclear industry, which can at a later date be made available to U.S. and European markets, when we do return to the nuclear option.

Over the past decade, operational safety of nuclear plants in OECD countries has been excellent. We need to continue this strong record of high safety standards. But maintaining our high standards of safety is not good enough. We need to ensure, to the best of our ability, that all countries operating nuclear plants meet high safety standards.

Safety of commercial nuclear power plants in the former Soviet Union is of primary concern. Another Chernobyl-scale accident would seriously endanger the future of nuclear energy world-wide. Last April, the G-7 held a special Nuclear Safety Summit in Moscow. The Summit recognized the need for safety and security-related technical (and possible financial) assistance.

Plans to develop and/or expand nuclear energy programs in China, South Korea, Taiwan, and other rapidly industrializing countries need to be monitored carefully. A culture of safety and accountability, as is present in OECD countries, needs to be

---

[15] This is not the case with any of the fossil fuels. We do not include the cost of the externalities of fossil fuel use (whether they be national security or environmental) in our price calculations.

impressed upon these countries. This is a role which Japan, which I believe has the most advanced commercial nuclear energy program in the world, needs to play. Japan has an outstanding record in safety and accountability. It should take the lead in promoting these same standards among its Asian neighbors.

Non-proliferation concerns will have a direct impact on nuclear energy's future. Non-proliferation challenges are much more varied and complex than they were during the height of the Cold War—a time when two nations stood poised with nuclear weapons ready to destroy one another. Today's threat from nuclear weapons is more complex and much more difficult to identify. While the United States and Russia maintain their nuclear arsenals (along with the other declared nuclear weapon states), other countries, such as Iran, North Korea, and Iraq seek to develop nuclear weapons.

What can be done to reduce the threat of the proliferation of nuclear weapons? First, the role of the International Atomic Energy Agency (IAEA) needs to be strengthened. The IAEA has played a vital role in reducing the danger of nuclear weapons development in Iraq and North Korea. Other challenges await. We need to ensure that it has the proper resources to carry out its monitoring functions.

Second, we need to continue our close cooperation with our Japanese and European colleagues. The foundation of this cooperation is both the U.S.-Japan Nuclear Cooperation Agreement and the U.S.-EURATOM Agreement. I applaud Ambassador Kennedy's efforts in negotiating the U.S.-Japan agreement and was delighted to see the U.S.-EURATOM agreement recently come into force. We need to work closely with our European and Japanese allies. As part of this partnership, we should respect and support their decisions to reprocess spent nuclear fuel. Japan and Europe view reprocessing as an integral part of their national energy policies. Instead of worrying about their reprocessing programs, we should be working with them in ensuring the safe and proliferation-resistant development of nuclear power in other parts of the world.

These are tough challenges, but let me say that they are no tougher than the national security, energy security and environmental problems associated with other alternatives.

Successful Nuclear Energy Programs: Japan

I do not only want to focus on the challenges of nuclear power, but also to look at its successes and opportunities. Having worked closely with Ambassador Imai over the past year, I have learned much about the Japanese nuclear energy program. I know he was invited to speak at this weekend's conference, but was unable to attend. So in his absence, I would like to share some brief thoughts on what I learned from Ambassador Imai about one of the world's most advanced commercial nuclear power programs.

As a national security advocate, I understand and support the reasoning behind Japan's decision to make nuclear energy one of its principal energy resources in the next century. It is the world's second largest economy, yet has no indigenous energy resources. Oil, which accounts for over 50 percent of its primary energy supply, is almost 100 percent imported—70 percent of which comes from the Persian Gulf. Energy is taken very seriously in Japan and given top priority.

In typical Japanese fashion, they are undertaking a long-term nuclear energy strategy—with the goal of developing a fully closed nuclear fuel cycle. This long-term strategy is being carried out with future generations in mind, with the objective of securing for them a safe, sustainable, reliable, and environmentally sensitive fuel source in the twenty-first century.

This program includes the reprocessing of spent fuel to recover uranium and plutonium to be reused in existing commercial light water reactors in the form of MOX fuel and continuing development of fast breeder reactor technology. Fast breeder reactor technology is expected to become commercially viable around 2030.

As with any new technology, the costs are high, but so is the cost of maintaining a military presence in the Gulf region to secure oil for America and the world. Japan's long-term nuclear energy program is its contribution to maintaining our global energy security.

Medium to Long-Term Nuclear Energy Technologies

Discussion of Japan's long-term nuclear energy program brings me to the issue of continuing our development of a number of advanced nuclear technologies—a discussion that is best left to the technical experts at this conference. Nevertheless, I still would like to make several brief comments from a layperson's perspective.

If nuclear power is to play a dominant role in the 21st century, research and development of advanced nuclear reactors and technologies must continue—just as is the case with any other energy technology. One such advanced nuclear technology under development by the French and Japanese is reprocessing, whereby uranium and plutonium are recovered from spent fuel and can be reused in existing commercial light water reactors in the form of mixed-oxide fuel (MOX). I understand the French are already using MOX fuel in several of their commercial light water reactors and the Japanese have plans to do so toward the end of the decade. I think reprocessing is a viable technological option and one that makes economic sense over the medium to long-term.

Although uranium resources are currently abundant and inexpensive, which raises questions about the economics of reprocessing and long-term research on the breeder reactor technology, the situation could change. If we are to achieve our sustainable growth scenario—whereby North America and OECD Europe return to nuclear energy for reasons of economic, national and environmental security reprocessing will not be a luxury, but a necessity.

Part of our GEMS Model includes a relatively straight forward nuclear fuel cycle analysis, which can project natural uranium civilian reactor requirements out to 2050 and compare these figures with current known uranium resources.

Figure 8 compares the current known supply of natural uranium to aggregate consumption by civilian nuclear power reactors. The current known supply of uranium recoverable for under $80 per kilogram is 3,800,000 tons according to the Uranium Institute. Total natural uranium resources minable for under $130 per kilogram is 6,200,00 tons. Cumulative demand for uranium projected by the GEMS Model show that the cheaper sources of natural uranium (under $80 per kilogram) will be exhausted by the year 2045 in the business as usual scenario, and by 2036 in the sustainable growth scenario. After this time, the burning of reprocessed uranium in the form of MOX fuel will become cost efficient. More noticeably, the increased use of nuclear fuel in the sustainable growth scenario results in a rapid increase in uranium prices after the year 2030, approaching $130 per kilogram by the year 2050.

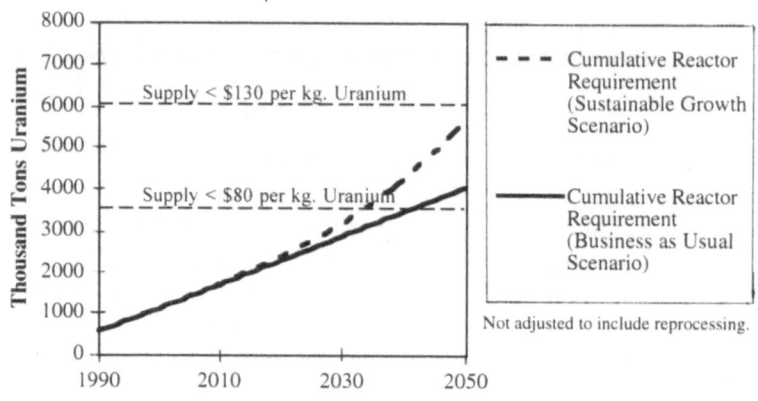

Source: GEMS Global Energy Supply and Demand Model

Figure 8. Cumulative Civilian Reactor Uranium Demand Compared to Known Uranium Resources

PART V: CONCLUSIONS

This morning I provided you with a brief overview of our energy future—both in the short-term (the next 15 years) and the long-term (the next 50 years).

I have attempted to frame this analysis within the concept of sustainable development—a definition I have expanded to include both national security and environmental quality concerns. In carrying out this examination, I have also paid close attention to nuclear power—assessing its current contribution, the challenges it faces, and the important role it can play in the twenty-first century.

The first part of my presentation focused on what I believe is one of the most serious national security threats we face within the next 15 years—our increasing reliance on oil from the Persian Gulf.

The second part of my presentation focused on energy supply and demand patterns over the short-term (the next 15 years) and the huge increase in energy demand coming from the rapidly developing countries. Yet despite our commitments under Rio and pledges by many governments to achieve sustainable development, we continue our trend toward greater reliance on fossil fuels. This trend is clearly not a realistic one if we are to achieve a sustainable future.

The third part of my presentation focused on energy supply and demand patterns over the long-term (the next 50 years). In assessing our long-term future, I presented two case scenarios. The business as usual scenario—in which we have a continuation of present policies and market conditions—is clearly not sustainable. The sustainable growth scenario—which includes better energy efficiency, more conservation, reduction in oil and coal use, expanded natural gas, greater market penetration by renewables, and a slight increase in nuclear's share of electricity generation—provides us with a more sustainable future. We see, however, that if we are to achieve a sustainable future, we have hard work ahead of us. If we are to achieve our goal of sustainable development, there will be no silver bullet that will solve our energy problems. Instead, our solution will be found through reliance on a variety of energy technologies, including

- Renewables: ensuring they become a serious player in energy markets.
- Energy efficiency and conservation: continuing the trend toward improving energy efficiency and energy conservation.
- Natural gas: expanding the natural gas infrastructure, which will require unprecedented cooperation and capital in building trans-continental pipelines.
- Nuclear: maintaining the nuclear option, which will require advanced nuclear technologies that are cost effective and safe.

The fourth part of my presentation focused on what I termed the short-term challenges of nuclear power, but also the successes and long-term technological options. Short-term challenges include waste disposal, economics, safety, and non-proliferation. These are tough challenges, but let me say that they are no tougher than the national security, energy security and environmental problems associated with other alternatives. Long-term technological options include improving reprocessing and eventual commercialization of fast breeder reactor technology.

Finally, I would like to touch upon what important messages a conference like this should be sending to key policymakers in the United States, OECD countries, and the world. What needs to be said about nuclear?

Policymakers in many countries seem to be looking primarily at the short-term problems of nuclear, and not the long-term solutions it can provide. That is why it is so important to look at the global dynamics and the long-term in order to assess the costs and the benefit of pursuing more aggressively the nuclear energy future.

The development of nuclear fission was driven by a national security need. Great expense and great effort went into development of the nuclear deterrence option—not because it was economical, but because it was a national security necessity.

But unless there is an immediate threat, as was the case in World War II, awareness and progress come slowly. I envision a five part strategy to more fully explore revitalization of the nuclear option:

- Promoting understanding of energy supply and demand in the next century and its impact on national security, economics, and the environment. I would hope that reports like that of the Trilateral Commission help in this process.

- Examining nuclear futures. I applaud the effort of Los Alamos National Laboratory: Sig Hecker, Ed Arthur, Rich Wagner and others who are tackling this problem and attempting to lay out options and pathways the future. If we do not have such a plan, how can we know where we are going?

- Facilitating discussions within the Administration and Congress on long-term energy futures. More discussion on the costs and benefits of the "business as usual" versus "sustainable growth" scenario. Two recent reports by this Administration, *Science and National Security* and the report of the President's Council for Sustainable Development illustrate that there is opportunity, in my view, to discuss nuclear energy's contribution to resolution of important environmental challenges, such as global climate change.

- Maintaining our nuclear capability and infrastructure in this country. This includes: supporting our universities and national laboratories in continuing research and development of nuclear technologies; ensuring that our domestic industry is able to compete in East Asia as that region begins a large expansion of its nuclear generating capacity; and taking our place at the nuclear table on the global level—whether the discussion concerns proliferation or commercial topics. Who do we want selling nuclear technology to the Chinese—Russia and North Korea or GE and Westinghouse?

- Continuing close consultation with the international community through IEA, the IAEA, NEA and bilaterally, especially with Europe, Japan and Russia. And this should not be limited to government to government talks—I applaud the effort of Chatham House and the Council on Foreign Relations to look at nuclear futures: problems and potential.

This morning I have tried to "back into" the subject of nuclear power from an overall national security, environment and economic perspective. Often, we focus on the problems of "kick-starting" the nuclear option. But my suggestion is to make nuclear power an element of national security and sustainable development. Nuclear energy's roots were in national security. Its future may lie with sustainable development. And it is here that this option must be best understood. Every energy option has problems, but the world will have a larger problem if energy needs for our expanding globe are not met.

APPENDIX A: DESCRIPTION OF GEMS GLOBAL ENERGY SUPPLY AND DEMAND MODEL

The GEMS Model compiles energy statistics from the base year 1990[16] to develop energy demand projections for Canada, China, France, Germany, India, Japan, South Korea, Taiwan, the United Kingdom, and the United States. These projections are incorporated into the regional projections compiled for Africa, the former Soviet Union, Latin America, the Middle East, Non-OECD Asia, Non-OECD Europe, OECD Europe, OECD North America, OECD Pacific and the OECD Total. The regional totals are aggregated in the world total. GEMS uses the standard energy supply/demand integration sheet used extensively by energy analysts around the world.

For this presentation, GEMS has developed two scenario cases: (1) "business as usual" and (2) "sustainable growth."[17] In the business as usual scenario, projections are based on current practices that rely heavily on oil for transport and coal for electrification—in other words, a continuation of present policies. It also accounts for continued progress in energy efficiency in the transport, industry, and residential sectors, and growth in use of natural gas, including development of long distance pipelines. It assumes that renewables make no serious market penetration. Nuclear energy generating capacity declines in North America and OECD Europe, but increases in East Asia. In the sustainable growth scenario, we assume vast improvements in energy conservation and efficiency; growth in the role of renewables in the electricity generation sector, and augmentation of natural gas use in the transport, industrial, commercial, and industrial sectors. Nuclear energy's role is slightly increased from today's level of generating approximately 17 percent of the world's electricity to generating about 20 percent of the world's electricity. We also assume that electric vehicles account for 30 percent of the market share in the transportation sector by 2050.

An important assumption of the GEMS model is the projected economic growth of each region and country through 2050. The economic growth assumptions are calculated from data available from the OECD Secretariat and the World Bank. The GEMS model, however, is generally more realistic about world economic growth than either of these two organizations. The average world economic growth from 1990 to 2050 estimated by the GEMS Model is 2.4 percent. Population growth will be an important determining factor for economic growth rates—especially the increasing trend of urbanization. Our estimates for population growth over the next 50 years are 1.7 percent,

---

[16] All projections are made from base-year statistics 1990, compiled from IEA publications: *Energy Statistics and Balances of Non-OECD Countries 1989-1990* (Paris) and *Energy Balances of OECD Countries 1990-1991* (Paris).

[17] The final energy demand total in 2050 under the sustainable growth scenario (approximately 330 million barrels per day oil equivalent) is lower than the final energy demand total under business as usual (approximately 380 million barrels per day oil equivalent) because of the assumed increases in energy efficiency and conservation under the sustainable growth case.

which is considerably lower than the average growth rate of 2.1 percent from 1971 to 1992.

The energy /GDP ratio is one of the most important assumptions made in the model. This ratio determines the energy intensity of an economy and represents an important distinction between rapidly industrializing (China and India) and industrialized (OECD) countries. Rapidly industrializing countries have a higher energy use per unit of GDP because their industries (steel and heavy manufacturing) are more energy-intensive. Industrialized countries have lower energy /GDP ratios because their industries (service industries) are less energy-intensive. This ratio is multiplied by the economic growth rate to determine the final energy demand growth rate. The GEMS Model takes into account that the energy efficiency of technologies used by developing countries is steadily improving. While they may not always adopt the most efficient technologies, the basic standard has risen dramatically over the last 15 to 20 years, allowing developing countries to leap-frog to more efficient energy technologies. Having the benefit of more efficient technologies will result in developing countries having lower energy/GDP ratios than currently industrialized countries at a similar earlier stage of development. The final demand growth rate is found by multiplying the estimated economic growth figure by the energy /GDP ratio. It is used to determine the growth of final energy demand for each region and country in GEMS. The sector demand shares by fuel for the transportation, industrial and commercial/residential sectors, are based on the 1990 IEA data. For each country and region, depending on the scenario, changes are made to the sector demand shares to show increased (or decreased) usage of oil, coal, natural gas, nuclear, and renewable energy sources in each sector over the 60 year period. Changes in the relative importance of the transportation, industrial, and commercial/residential sectors also illustrate changing energy needs for the countries and regions in the GEMS model. Figure 9 illustrates the GEMS Model calculation methodology.

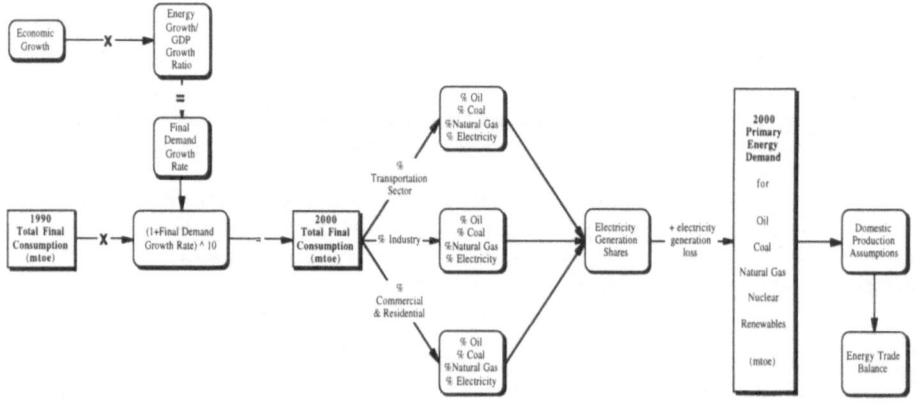

Figure 9. GEMS Model Calculation Methodology

WORLD

CASE SCENARIO: BUSINESS AS USUAL

Growth Assumptions

GROWTH ASSUMPTIONS	1990-2000	2000-2010	2010-2025	2025-2050
ECONOMIC GROWTH % [A2]	2.45	2.50	2.40	2.40
ENERGY/GDP RATIO % [A3]	0.70	0.70	0.60	0.60
FINAL DEMAND GROWTH RATE % [A4]	1.72	1.75	1.44	1.44

Supply Assumptions

SUPPLY ASSUMPTIONS	1990	2000	2010	2025	2050
INDIGENOUS PRODUCTION (millions metric tons of oil equivalent)					
COAL [A9]	2322.02	2716.39	3459.44	4497.14	6277.54
PETROLEUM [A10]	3236.25	3205.42	3800.47	4797.88	6995.88
GAS [A11]	1702.45	1603.11	1901.38	2368.71	3293.82
NUCLEAR [A12]	515.39	521.35	545.75	654.29	670.71
RENEWABLES [A13]	248.03	336.54	468.44	525.22	800.66
NUCLEAR CAPACITY (GWe) [A14]	325.83	329.60	345.03	361.94	349.20
EXPORTS/IMPORTS (NET) [A15]					
COAL [A16]	-19.06	0.00	0.00	0.00	0.00
PETROLEUM [A17]	-175.60	0.00	0.00	0.00	0.00
GAS [A18]	-21.76	0.00	0.00	0.00	0.00
ELECTRICITY [A19]	-0.60	0.00	0.00	0.00	0.00
SECTOR DEMAND SHARES %					
TRANSPORTATION [A22]	27.45	28.00	28.00	28.00	28.00
INDUSTRIAL [A23]	47.15	47.92	47.95	48.49	47.35
COMMERCIAL/RESIDENTIAL [A24]	25.40	24.08	24.05	23.51	24.65

Fuel Mix and Generation Shares

	1990 [I1]	2000 [I1]	2010 [J1]	2025 [K1]	2050 [L1]
TRANSPORTATION FUEL MIX %					
COAL [G3]	1.37	1.52	1.56	1.58	1.52
PETROLEUM [G4]	96.63	96.49	95.29	94.28	92.16
GAS [G5]	0.67	0.56	0.57	0.57	0.60
ELECTRICITY [G6]	1.34	1.42	2.58	3.57	5.72
INDUSTRIAL FUEL MIX %					
COAL [G9]	34.04	32.03	32.60	32.94	33.06
PETROLEUM [G10]	22.79	31.40	30.92	30.72	30.79
GAS [G11]	24.43	19.64	18.71	17.80	16.47
ELECTRICITY [G12]	18.74	16.92	17.78	18.54	19.68
COMMERCIAL/RESIDENTIAL FUEL MIX %					
COAL [G15]	18.36	18.80	18.39	17.55	16.50
PETROLEUM [G16]	20.49	21.14	21.42	22.49	24.36
GAS [G17]	31.38	29.73	28.76	27.00	24.61
ELECTRICITY [G18]	29.77	30.32	31.44	32.97	34.54
ELECTRICAL GENERATION SHARE %					
COAL [G21]	39.82	43.44	45.59	46.66	46.08
PETROLEUM [G22]	15.87	14.57	14.17	15.02	17.14
GAS [G23]	20.80	18.13	17.94	18.49	19.23
NUCLEAR [G24]	15.87	14.50	12.00	11.00	8.00
RENEWABLES [G25]	7.64	9.36	10.30	8.83	9.55

WORLD

GROWTH ASSUMPTIONS	1990-2000	2000-2010	2010-2025	2025-2050
ECONOMIC GROWTH % [A2]	2.45	2.50	2.40	2.40
ENERGY/GDP RATIO % [A3]	0.70	0.70	0.50	0.50
FINAL DEMAND GROWTH RATE % [A4]	1.72	1.75	1.20	1.20

SUPPLY ASSUMPTIONS	1990	2000	2010	2025	2050
INDIGENOUS PRODUCTION					
(millions metric tons of oil equivalent)					
COAL [A9]	2322.02	2274.30	2640.48	2977.95	3740.47
PETROLEUM [A10]	3236.25	2938.16	3438.07	3569.97	3925.42
GAS [A11]	1702.45	1697.27	1930.75	2418.81	4446.81
NUCLEAR [A12]	515.39	522.69	526.60	694.59	1515.86
RENEWABLES [A13]	248.03	137.88	150.46	255.90	1061.11
NUCLEAR CAPACITY (GWe) [A14]	325.83	330.45	332.92	439.12	958.33
EXPORTS/IMPORTS (NET) [A15]					
COAL [A16]	-19.06	0.00	0.00	0.00	0.00
PETROLEUM [A17]	-175.60	0.00	0.00	0.00	0.00
GAS [A18]	-21.76	0.00	0.00	0.00	0.00
ELECTRICITY [A19]	-0.60	0.00	0.00	0.00	0.00
SECTOR DEMAND SHARES %					
TRANSPORTATION [A22]	27.45	28.00	28.00	28.00	28.00
INDUSTRIAL [A23]	47.15	47.92	47.85	48.03	48.65
COMMERCIAL/RESIDENTIAL [A24]	25.40	24.08	24.15	23.97	23.35

	1990 [H]	2000 [I]	2010 [J1]	2025 [K1]	2050 [L1]
TRANSPORTATION FUEL MIX %					
COAL [G3]	1.37	1.50	1.25	1.00	0.00
PETROLEUM [G4]	96.63	97.00	96.00	85.00	60.00
GAS [G5]	0.67	0.67	2.00	4.00	10.00
ELECTRICITY [G6]	1.34	0.83	3.50	10.00	30.00
INDUSTRIAL FUEL MIX %					
COAL [G9]	34.04	32.00	32.00	30.00	20.00
PETROLEUM [G10]	22.79	24.00	25.00	22.00	18.00
GAS [G11]	24.43	25.00	25.00	25.00	27.00
ELECTRICITY [G12]	18.74	19.00	20.00	23.00	35.00
COMMERCIAL/RESIDENTIAL FUEL MIX %					
COAL [G15]	18.36	18.50	18.50	15.00	10.00
PETROLEUM [G16]	20.49	21.00	22.00	20.00	15.00
GAS [G17]	31.38	30.50	29.50	28.50	35.00
ELECTRICITY [G18]	29.77	30.00	30.00	36.50	40.00
ELECTRICAL GENERATION SHARE %					
COAL [G21]	39.82	39.82	42.50	41.00	32.00
PETROLEUM [G22]	15.87	16.00	15.00	11.00	10.00
GAS [G23]	20.80	20.80	20.00	22.00	24.00
NUCLEAR [G24]	15.87	18.50	17.50	19.00	20.00
RENEWABLES [G25]	7.64	4.88	5.00	7.00	14.00

CASE SCENARIO: SUSTAINABLE GROWTH

REFERENCES

Atomic Energy Commission of Japan. (1994) *Long-Term Program for Research, Development, and Utilization of Nuclear Energy.* Tokyo: Atomic Energy Commission of Japan.

Boiteux, M. (1995) *Concurrence, Regulation, Service Publique.* Electricité de France.

Bond, G. and L. Carter. (1994) *Financing Private Infrastructure Projects: Emerging Trends from IFC's Experience.* Discussion Paper 23. Washington, DC: World Bank.

Carnegie Endowment for International Peace and the International House of Japan. (1995) *The United States, Japan, and the Future of Nuclear Weapons.* [Report of the U.S.-Japan Study Group on Arms Control and Non-Proliferation After the Cold War. Co-Chairman: William Clark, Jr. and Ryukichi Imai.] Washington, DC: Carnegie Endowment for International Peace.

Center for Strategic and International Studies. (1995) *Nuclear Energy Safety Challenges in the Former Soviet Union.* [A Consensus Report of the CSIS Congressional Study Group on Nuclear Energy Safety Challenges in the Former Soviet Union -- Project Chairman: The Honorable James Schlesinger] Washington, DC: CSIS.

Department of Energy/Energy Information Agency. (1995) *Emissions Greenhouse Gases in the United States: 1987-1994.* Washington, DC: United States Government Printing Office.

Department of Energy/Energy Information Agency. (1995) *Sustainable Energy Strategy: Clean and Secure Energy for a Competitive Economy.* Washington, DC: U.S. Government Printing Office.

Department of Energy/Energy Information Agency. (1995) *World Nuclear Outlook 1995.* Washington, DC: United States Government Printing Office.

Department of Trade and Industry. (1995) *The Energy Report: Competition, Competitiveness, and Sustainability.* Volume I.

Dunkerley, J. and D. Jhirad. (1995) Editors. Special Issue: Financing the Energy Sector in Developing Countries. *Energy Policy.* Volume 23. Number 11.

European Commission. (1995) *An Energy Policy for the European Union.*
Brussels: European Union.

Foell, W., M. Amann, G. Carmichael, M. Chadwick, J.P. Hettelingh, L. Hordijk, and
Z. Dianwu. Eds. (1995) *Rains-Asia: An Assessment Model for Air Pollution in
Asia.* Report on the World Bank Sponsored Project: "Acid Rain and Emission
Reduction in Asia." RAINS-ASIA Project. Washington, DC: World Bank.

Imai, R. (1995) *A Call for Regional Nuclear Cooperation. Japan Review of
International Affairs.*

Imai, R. (1995) *Europe and Asia: Working Together to Strengthen Non-
Proliferation and Promote Disarmament.* Paper presented to the Japanese-
German Center, Berlin.

Imai, R. (1995) Fifty Years After Hiroshima: A Half Century of Nuclear Energy
and Beyond. *Institute for International Policy Studies.* IIPS Policy Paper 142E.

Imai, R. (1993) "The Long Shadow of Nuclear Weapons," *International
Institute of Global Peace.* IIGP Policy Paper 144E.

Intergovernmental Panel on Climate Change. (1995) *Climate Change 1995:
The Science of Climate Change.* [Draft Summary for Policy-makers of the draft
contribution of Working Group I to the IPCC Second Assessment Report.]
World Meteorological Organization and United Nations Environment Programme.

Intergovernmental Panel on Climate Change. (1990) *Climate Change: The
IPCC Scientific Assessment.* Cambridge: Cambridge University Press.

International Energy Agency. (1995) *The Energy Charter Treaty: A
Description of Its Provisions.* Paris: OECD.

International Energy Agency. (1994) *Energy in Developing Countries: A
Sectoral Analysis.* Paris: OECD.

International Energy Agency. (1995) *Energy Policies of IEA Countries: 1994
Review.* Paris: OECD.

International Energy Agency. (1995) *Energy Policies of the Russian
Federation.* Paris: OECD.

International Energy Agency. (1995) *North Sea Oil Supply: The Expected
Peak Recedes Again.* Paris: OECD.

International Energy Agency. (1995) *Oil, Gas, and Coal Supply Outlook*. Paris: OECD.

International Energy Agency. (1996) *World Energy Outlook*. Paris: OECD.

Kozloff, K.L. and R.C. Dower. (1993) *A New Power Base: Renewable Policies for the Nineties and Beyond*. Washington, DC: World Resources Institute.

Martin, W.F. H. Steeg. and R. Imai. (1996) *Maintaining Energy Security in a Global Context*. New York: The Trilateral Commission.

Ministry of International Trade and Industry. (1995) *Energy in Japan: Facts and Figures*. Tokyo: Ministry of International Trade and Industry.

Ministry of International Trade and Industry. (1994) *Japan's Long-Term Energy Plan*. Tokyo: Ministry of International Trade and Industry.

National Academy of Sciences. (1994) *Management and Disposition of Excess Weapons Plutonium*. Washington, DC: National Academy Press.

National Academy of Sciences. (1995) *Management and Disposition of Excess Weapons Plutonium: Reactor-Related Options*. Washington, DC: National Academy Press.

National Oceanic and Atmospheric Administration et al. (1995) *Scientific Assessment of Ozone Depletion*. Geneva: World Meteorological Organization.

Natural Resources Canada. (1994) *Canada's Energy Outlook: 1992-2010*. Ottawa: Ministry of Supply and Services.

Nuclear Energy Agency. (1995) *Nuclear Energy Data*. Paris: OECD.

Nuclear Energy Agency. (1995) *Nuclear Energy Programmes in OECD/NEA Countries*. Paris: OECD.

Office of the Secretary of Defense. (1996) *Proliferation: Threat and Response*, Washington, DC: U.S. Department of Defense.

Organization for Economic Cooperation and Development. (1995) *Global Warming: Economic Dimensions and Policy Responses*. Paris: OECD.

President's Council on Sustainable Development. (1996) *Sustainable America: A New Consensus*. Washington, DC: U.S. Government Printing Office.

United Nations Environment Programme. (1992) *Report of the Fourth Meeting of the Parties to the Montreal Protocol on Substances that Deplete the Ozone Layer.* Copenhagen: UNEP.

World Bank. (1993) *Energy Efficiency and Conservation in the Developing World: The World Bank's Role.* A World Bank Policy Paper. Washington, DC.

World Bank. (1993) *World Bank's Role in the Electric Power Sector: Policies for Effective Institutional, Regulatory, and Financial Reform.* A World Bank Policy Paper. Washington, DC.

World Bank. (1992) *World Development Report 1992: Development and the Environment.* New York: Oxford University Press.

World Bank/International Finance Corporation. (1994) *Power and Energy Efficiency Status Report on the Bank's Policy and IFC's Activities.* Joint World Bank/IFC Seminar Presented to the Executive Directors.

World Commission on Environment and Development. (1987) *Our Common Future.* Oxford: Oxford University Press.

World Health Organization and United Nations Environment Programme. (1992) *Urban Air Pollution in the Megacities of the World.*

World Resources Institute. (1996) *World Resources 1996-97: A Guide to the Global Environment.* New York: Oxford University Press.

Zhaobo, C. (1995) *The Development and Prospects of Nuclear Energy in China.* Paper Presented to the Sixth International Conference on Nuclear Cooperation

BACKGROUND ON WILLIAM F. MARTIN

William Martin served as U.S. Deputy Secretary of Energy, Executive Secretary of the National Security Council in the White House, and Special Assistant to President Ronald Reagan. In 1992, he was Executive Director of the Republican Party National Platform Committee.

During the second oil crisis, Mr. Martin was Special Assistant to the Executive Director of the International Energy Agency, Ulf Lantzke. He also held positions in the Department of State and MIT's Workshop on Alternative Energy Strategies headed by Carroll L. Wilson.

Today, he is Chairman of Washington Policy and Analysis, Inc., an international energy consulting firm. He holds degrees from the University of Pennsylvania and MIT. Mr. Martin is a member of the Board of the World Resources Institute and a member of the Council on Foreign Relations, New York.

He was recently the lead author of a Trilateral Commission study, *Maintaining Energy Security in a Global Context.* (September 1996)

THE NUCLEAR FUEL CYCLE AND OTHER FEATURES OF NUCLEAR POWER: REACHING A PUBLIC CONSENSUS

Richard Wilson

Department of Physics
Harvard University
Cambridge, MA 02138

INTRODUCTION

Fermi's Dream and Ford Foundation's Nightmare.

Before 1975 there were coherent and widely accepted plans for a nuclear fuel cycle in the world which for brevity I will call Fermi's dream. One starts from uranium ore, processing it to form uranium metal, burning the uranium 235 in an electric power producing reactor, reprocessing the fuel to separate the uranium 235 and plutonium 239 for use in subsequent reactors. If the reactor is a fast neutron reactor, other transuranic elements can also be broken up and destroyed by fission. All that would be left for subsequent waste disposal would be fission products themselves, almost all with half lives of 30 years or less. The fast neutron reactor of preference was cooled with liquid sodium. With this it was envisaged that all the energy in uranium, including both isotopes could be unlocked.

In the early 1970s problems appeared in the Fermi dream. It was realized that the existence of many tons of chemically separated plutonium might lead to a possibility of theft, or "diversion", of enough material to make a nuclear bomb. The presence in the hands of a small "rogue" country, or of a terrorist group, is unacceptable and would be a nightmare. This led a study sponsored by the Ford Foundation (Keeny et al., 1977) and the decision of President Carter on April 7th 1977 to abandon the plans in the U.S. to reprocess spent nuclear fuel, and slow the development of the breeder reactor. Since that time the U.S. has split into two groups that talk past each other instead of to each other. They have taken firm positions and will not compromise, so that now instead of a generally accepted policy the U.S. is divided. Alas this has not prevented the U.S. administration from lecturing other countries about the dangers of reprocessing.

On the one hand, Keeny et al. argued that by refraining from reprocessing the U.S was setting an example to the rest of the world and indicating that the U.S. believed that a breeder reactor is not necessary. President Carter took this argument and made it a policy.

Technology for Global Economic and Environmental Survival and Prosperity
Edited by Kursunoglu *et al.*, Plenum Press, New York, 1997

67

In the present U.S Administration this has been extended by some officials into a statement that we should not burn plutonium in reactors - again in order to set an example. On the other hand there are a few in the U.S. who suggest that the US should immediately begin to reprocess fuel again.

This subject interested Andrei Dmitreyvich Sakharov while he was alive. He was an optimist, yet a realist. He insisted that human rights without human (technological) progress was not possible, since freedom from want could not be achieved. Yet technological progress without attention to human rights and liberty leads to a failure to manage the technology, with disasters such as occurred at Chornobyl and Bhopal. Andrei Sakharov also argued that on technological matters of public importance scientists without a government position should discuss the issues and recommend actions.

WHAT ARE THE REASONS FOR WANTING NUCLEAR POWER?

There are three basic reasons why environmentalists should be strongly interested in advancing nuclear power:

(1)   The major alternate to nuclear power is the burning of coal which inevitably produces air pollution. Even after the controls of the last century, many scientists believe that fine (probably acid) particles cause premature death for 70,000 people a year in the USA and proportionately elsewhere (Shprentz, Bryner and Shprentz 1996, Wilson and Spengler 1996). Nuclear power in ordinary operation has no important pollutant emissions.

(2)   The burning of any fossil fuel produces carbon dioxide, which probably changes the earth's greenhouse, and may induce major climate change. This has been advertised by the present Vice President of the U.S. A. (Gore, 1990). At the Rio de Janeiro in 1991 and in Berlin in 1995 the world's politicians agreed to cap the greenhouse gas emissions and study further reductions in emissions. Three leading scientists with contrasting opinions on the reliability of the scientific data, nonetheless called for use of alternates to fossil fuels when economically feasible (Singer, Revelle, and Starr 1991).

(3)   One long-term requirement of any energy system is its sustainability. It is important that one has enough fuel for the foreseeable future. Fermi's dream of a breeder reactor clearly encompassed this goal. Since we burn up 100 times as much uranium as in a light water reactor, we can afford to pay more for the raw uranium and have enough at modest cost for 100,000 years. No other energy source presently can guarantee energy supplies at a known modest cost for a long time. Pietr Kapitza and Sakharov both discussed this (Kapitza, 1970; Sakharov, 1978), and Sakharov discussed it again at the historic Conference on a Nuclear Free World in Moscow in February 1987.

The first set of questions that I ask is:

How well known are these three reasons, how well are they accepted, acknowledged and what priority is attached to them? (particularly by those opposed to expansion of nuclear electric power)

These reasons have not yet been driving forces for expansion of the role of nuclear powered electricity. Even though the role of nuclear electric power is still expanding in some countries, it seems that other negative thoughts about nuclear power overshadow these three reasons in many people's minds, and in the "trade-off" which each individual, organization and government must make, nuclear electric power has suffered in many other countries. In particular, there many people seem to hope that alternative energy sources to fossil fuel and nuclear power will become economically viable in the next decade or two. Therefore it is vital to understand what these other negative thoughts are, and to determine whether and how they can be made to be less significant.

ALTERNATIVES FOR AN ENERGY FUTURE

Before 1973 in OECD countries, and in the USSR, the cost of fuels, and electricity, was continually declining in real terms. This led to what would now be termed wasteful practices, and what would now seem to be an excessive demand for new electricity generating stations. Since 1973, the total energy per capita has stayed constant or declined, although electricity use has continued to rise. This has led many analysts to believe that more efficient end use can lead to further reductions in fuel use in OECD countries, and can slow the rise in developing countries.

If one discounts, with a potential peril to society, the air pollution and the threat of global warming, there is an option for much of the world: to burn fossil fuels, and in particular coal, for the next several hundred years. This is the road that the Peoples Republic of China is choosing, although they are hoping to expand the nuclear role considerably. Air pollution can, indeed, be reduced by pollution controls, but the cheaper ones make the power plant less efficient and can accentuate global warming. A more attractive temporary option is to burn natural gas. Natural gas, when it is available, can be burned efficiently and with much less (but not zero) air pollution and it is more readily available than was thought 20 years ago.

Overall there has been a marked change in perception of nuclear power both by the experts and by the general public since 1973. In particular there is general agreement that there is time to develop and consider many options. A potential option can include alternates to both fossil fuels and nuclear power of which wind generators seem the most promising (1996 busbar cost of 5 cents per kWhr not allowing for the intermittent nature). Such alternates to both fossil fuels and nuclear power must be reliable, sustainable, and adequately cheap sources of electricity. This then leads to the question:

Is the possibility of alternates being developed in the next 20 years large enough that mankind can afford not to develop the nuclear option as a possibility?

Even if one accepts the aforementioned reasons for developing nuclear power, the increase in availability of uranium (and for a few years the burning of plutonium released from weapons stockpiles) can delay the need for using the full energy potential of the uranium fuel. The urgency in developing a commercial breeder reactor that was felt by many experts in 1970 can be replaced by a sense that there is time "to do the job right." But an urgent question still remains:

Are there reasonable scenarios that demand the development of the breeder reactor on a rapid time scale?

A concern of many experts in the 1970s was that nuclear electric power was expanding at an unprecedented rate for an energy technology, (as described for example in papers by Cesare Marchetti) and that a slower, more deliberate, pace would be beneficial. We can still ask:

What is an optimum pace for development of a breeder reactor technology and does that match reasonable scenarios for the demand?

The program for the fuel cycle then should be different from that envisaged in 1970, or even at the International Fuel Cycle Evaluation (INFCE) in 1980. There is time - perhaps as much as 40 years - to experiment and learn how to make the fuel cycle acceptable to a broader group of people. The arguments that nuclear power advocates should make to their peoples and their governments are also different. No longer should they argue an urgent need; but neither should they accept that there exists a *known* alternative to breeding nuclear fuel to satisfy the environmental demands. They can be more general: the cost of developing an option for the future that may eventually be discarded (if the optimists are correct) is not great compared with the cost to society of not having a viable option when the time comes. It is important to open up options for ones' grandchildren: not to close them off. Anything else is unsustainable. This demands that the world should continue to spend money on research and development on the breeder reactor option.

We may have 40 years to develop a nuclear power program that includes use of plutonium, that is acceptable to the people. But we do not now have a coherent plan on how to proceed, and I do not believe that we even have a clear statement of the problem that can be generally accepted. Thus I ask both the politicians and the "industry":

What has been accomplished to make an acceptable nuclear program in the 20 years since the Carter decision? Either by the "industry" or by the authors of the Ford Foundation report?

THE PREVENTION OF CLANDESTINE NUCLEAR BOMBS

Many scientists have argued that the only real issue about nuclear power is the connection with weapons (e.g. Cottrell, 1981; Flowers et al., 1975). While this may not be completely true, I assume for the purposes of this discussion that if this issue could be "resolved", the other issues will fade into insignificance.

Before 1990, when the cold war was in progress, the existence of enough weapons plutonium for over 30,000 atomic bombs made the presence of plutonium in electricity producing reactors reactor fuel seem a small problem to many. Yet the Keeny et al. echoed the feelings of several scientists and others that there were inadequate plans in the world to control the plutonium in the civil sector. This feeling has increased now that the cold war is over. The focus of a 1994 U.S. National Academy of Sciences report (Panofsky et al., 1994) was on the disposition of excess weapons plutonium, but much of the concern follows through to plutonium in nuclear electric power. This was made somewhat clearer in the 1995 report of the American Nuclear Society (Seaborg et al., 1995). The National Academy report called the existence of the excess weapons plutonium "A clear and present danger" to the United States - wording that, if accepted, would demand immediate action from the President. Although this urgency has not been accepted (excess weapons plutonium is clearly less of a danger than the cold war was), most analysts agree that coping with excess weapons plutonium is important. However, the delays in coping with the military problem by the U.S. Administration gives time for the nuclear power community to consider a long term plan for management of plutonium in the civilian nuclear industry. Both the NAS and the ANS reports also emphasized the urgent importance of a secure storage of militarily usable plutonium in an internationally verifiable way - a vital subject that is peripheral to this paper.

Was the NAS 1994 (Panofsky) committee right that there is a clear and present danger to the USA? If so how can we persuade the President to act? Is there a clear and present danger to other countries too? If so why are they still waiting for the U.S. to act?

One of the options for disposing of excess weapons plutonium and ensuring that it will not ever be used for weapons is to burn it in a reactor. This is the only way of actually destroying it, and has therefore a number of psychological advantages. While disagreements remain it is noteworthy that at several public meetings many persons normally opposed to nuclear electric power have agreed that burning plutonium is a good, or even the best, option. If that is decided, and it certainly seems to be the intent of the Russian government to pursue this route, there will be a quantity of plutonium to fuel reactors for a few years. When this is combined with the low price of uranium ore and the existence of an excess of highly enriched uranium, there will be enough fissionable material in the next few years to reduce the urgency for reprocessing, but there will be a shortage of fabrication facilities for mixed oxide fuel.

This then leads to the question:

Should the immediate aims of operating and soon to be operating reprocessing plants be redirected to development aimed at understanding how to make reprocessing cheaper, and safer? and how to explain that to the public?

Instinctively most of us recoil at the idea of burying material that is potentially useful - such as the plutonium with its energy content. The specter of a "plutonium mine" has been raised. After 300 years when most of the fission products will have decayed it would be possible to recover the plutonium without the dangerous barrier of radioactivity. Thus it remains very important to have available the alternate of destroying the plutonium and actinides. While this can be done to a considerable extent by recycle in a slow neutron Light Water Reactor, it is easier and more complete in a fast neutron reactor.

This immediately leads to the question:

What is the cost advantage, either in direct technical costs or in public perception, of removing the long lived material before burial of the nuclear waste?

One of the premises of the Keeny et al., report was that it was technically easy, and not very much more expensive, to dispose of whole fuel rods including the long lived actinides, and not merely the fission products. This leads to the questions:

Is this presumption technically true? and

Why are not the authors of the Keeny report acting to persuade the government to demonstrate this soon by burying fuel rods now?

Plutonium separated from spent fuel is now piling up in many countries. A significant bottleneck is in the limited fabrication facilities that are available for plutonium or mixed oxide fuel. Now that it has been recognized that it is possible to make an atomic bomb (even if less reliable or low yield) with "reactor grade" plutonium with 25% of the isotope plutonium 240, this emphasizes the importance of secure storage and makes it a problem for many more countries (such as Japan and Germany) than those with excess weapons plutonium. This immediately raises the all important questions:

What, if any, is the difference between the ease of using "reactor grade" plutonium and "weapons grade" plutonium to make an atomic bomb? and can this difference (if any) be translated into a lower cost for protection of reactor grade plutonium?

This leads to another question:

How does one define secure storage? what are the criteria?

One used to define it as being sufficiently secure that a terrorist, or agent of a "rogue" country, would find it easier to steal from a military complex. But now there is a real possibility of disarmament, that seems inadequate.

Can one compare usefully possible terrorism with nuclear bombs and terrorism with other material?

It is relatively easy to obtain even large quantities of other explosives. A billion tons a year of ammonium nitrate is used in the U.S., about half as easily available fertilizer. While terrorists used about a ton to try to blow up the World Trade Center and the federal office building in Oklahoma City, it may be worth noting that a string of barges carries 10 kilotons (with the explosive power of a Hiroshima bomb, and a supertanker could carry 300 kilotons (the explosive power of a hydrogen bomb). The accidental explosions at Oppau in Germany and in Texas City, were larger than the Hiroshima explosion, and are examples of what a terrorist might do with ammonium nitrate in the future.

If plutonium is separated from spent fuel the volume is reduced and storage facilities can be smaller. This leads to the question:

Is there an appreciable advantage in storing separated plutonium in a central facility over storing unreprocessed fuel rods in many, dispersed, locations?

PROLIFERATION RESISTANT FUEL CYCLES

It has been claimed, and it seems reasonable, that facilities for chemically purifying small amounts of spent fuel are simple and cheap *if* environmental and occupational safety rules are not followed - and terrorist groups are unlikely to feel a need to follow them. It is therefore important that all spent fuel be accounted for - presumably by counting of, and measuring fuel rods. While not suggesting that the issue be forgotten, I note that this must be done even for present nuclear power plants, and therefore is not an additional issue for expanding the fuel chain into a fuel cycle. Moreover, over the long term the world must decide what to do about the increasing inventory of plutonium in the spent fuel rods

It is generally agreed that the most important danger is chemically separated plutonium. Since the most important item for the long term (>50 years) future is breeding fissile fuel from the uranium 238, and not merely using the fissile material in light water reactor spent fuel, I will simplify the discussion below by considering a breeder reactor. This must use fast neutrons, and the usual heat transfer medium discussed is liquid metal (sodium).

The Ford Foundation study raised the scepter of the "plutonium economy", a naive parody of the plans of the time. A world was envisaged with hundreds of breeder reactors, all plutonium fueled, and many in small unstable countries. Chemically pure plutonium, separated from the spent fuel would become an article of commerce and perhaps traded on the commodities exchanges (maybe restricted to the signatories of the non-proliferation treaty). It would then be all to easy for a terrorist to obtain a little plutonium to make a bomb. This image never materialized, and it remains important to describe a viable nuclear fuel cycle which is much safer. Such a cycle might include restriction of the use of a breeder reactor or recycled plutonium, to a relatively small number of industrialized countries with international inspection

A principal task of the designer of a breeder reactor cycle must therefore be to prevent the plutonium being easily diverted to uses as a terrorist or military use. Unless this can be done to public satisfaction, the reactor system is unlikely to achieve public acceptance.

Although a fast neutron reactor does not by itself create a plutonium economy, there seems little point in developing reactors which are probably more expensive if breeding is not to be accomplished. Therefore the fast neutron reactor has become synonymous with the plutonium economy in many minds. It is a useful thought for professionals in the nuclear industry to bear in mind when thinking about the future.

The "ordinary" reprocessing facilities used by COGEMA and British Nuclear Fuels Services (BNFS) follow the PUREX scheme which was designed for the military purpose of producing pure plutonium for bombs. Inevitably there is chemically pure plutonium available in the fuel cycle, although a modification called CIVEX would reduce this. This the raises the all important question:

Are the facilities at COGEMA and BNFS adequately secure?

Can the time available before more reprocessing is needed be used either to modify the existing cycle, or to demonstrate more clearly that it is already adequately secure?

THE IFR FUEL CYCLE

The success of the group at the Argonne National Laboratory in obtaining a high burn up with metal fuel suggested that a pyroprocessing scheme could be used. The key step in the reprocessing is electrorefining. The disassembled and dissolved fuel elements are first electrorefined to separate bulk uranium on a solid cathode. The remaining uranium along with plutonium and other actinides, minor in quantity but major in proliferation resistance, are electrorefined by deposition onto a liquid cadmium cathode. The cathodes are removed from the refiner cell, the cadmium and occluded salts are removed by retorting, and the uranium and uranium-plutonium actinide product is consolidated by melting.

In this process the minor actinides (neptunium, americium, and curium) and a little uranium accompany the plutonium product stream. Because these elements have similar electro potentials, the process cannot be simply modified to separate them. Moreover the product carries enough radioactive fission products to necessitate remote handling of even refabricated fuel. Unauthorized access is almost impossible and any attempt would be easy to detect. Moreover the process is compact and seems to be favorable in modest sizes. Therefore the fuel facility can be located at the reactor site, reducing the risk (albeit small) of transporting the spent fuel and refabricated fuel.

Technological feasibility of the pyroprocessing and fuel fabrication has been demonstrated and a data base established to ensure its practicality. The EBR-II fuel cycle facility is being refurbished. It is now (summer 1996) processing spent fuel from EBR-II. It had been proposed, and accepted by DOE in previous administrations, to start a prototype demonstration of the entire fuel cycle. However, under rather firm, non technical, instructions from DOE, it is now planned to close EBR-II, and make its restart difficult by removing the sodium. Many technologies proceed without such an integrated demonstration of the complete fuel cycle. But as any engineer knows, a practical demonstration is often worth a thousand good calculations.

This then raises the important question:

Are the potential advantages of the IFR fuel cycle great enough to demand an international effort to demonstrate engineering feasibility of the full cycle?

If the answer to this is yes:

Can the U.S. program be revived, or will the leadership go elsewhere?

INTERNATIONAL COOPERATION

The issues that are being discussed at this conference are not the parochial issues of one country, even one as large as the USA or Russia. They are issues that affect the whole world. It is therefore appropriate that the whole world consider them together. The research and development has turned out to be very costly and even the U.S. is reducing its outlays in this direction. It is therefore important to use all available facilities in the world in a cooperative manner.

Fortunately the end of the cold war makes cooperation between Russia and the USA easier. There has been progress. In summer 1994 fast neutron reactor experts from the Argonne National Laboratory were actively discouraged (again by the U.S. Administration) from discussing problems with their counterparts at Obninsk. This has changed. Now Argonne scientists are being informed, for example, of the detail of the fuel loadings at Beloyarsk.

This raises then the following question:

What are the possible ways of using existing internationally available, (or planned) facilities in such a research effort? e.g: Introducing an IFR fuel cycle into Beloyarsk, Monju and/or Phoenix? Modifying Le Hague or Sellafield to become even more proliferation resistant and/or cheaper to operate?

ECONOMIC ISSUES

It is clear from the public reaction to terrorist activities on aeroplanes that the public is willing to spend a lot more money to combat terrorist activities than the economists suggest. The economists suggest a Willingness to Pay of about $4,000,000 per statistical life saved. This puts pressure on nuclear energy to be cost effective in other ways.

Therefore we must ask:

What are the true economic costs of reprocessing using the present PUREX process and can they be brought down?

Studies from OECD suggest that it is now more expensive to fabricate free plutonium fuel than to both buy and fabricate uranium fuel. Any cost advantage for reprocessing would therefore seem to come from a lower cost to dispose of reprocessed waste without transuranics than for ordinary waste. Technical estimates of this cost advantage suggest that it is small. But this is an issue where psychological reasons dominate over technical ones. This leads to another question:

What are the advantages (expressed in $$$) of final disposal of fuel from which the plutonium and transuranics have been removed? including the advantages in public perception?

Until a waste policy exists this has to be a technical calculation only leaving out factors of public perception. This would probably give a very small advantage. But somehow we must figure out what advantages there may be in public perception, or leave this as a variable to be inserted at the last moment.

In this connection it is worth noting that Keeny et al. (Ford Foundation study) argued that spent fuel rods could be placed in a permanent repository almost as easily as the fuel without the recyclable transuranics. While this may be technically correct, in the intervening 20 years there has been no political progress in the U.S. in establishing such a repository.

Other economic questions include:

Is the IFR fuel cycle likely to be as cheap? cheaper? more expensive?

Is it economically preferable to store the excess pure plutonium waiting the 40+ years till a breeder reactor is needed (as the Russians presently plan), or to burn it in MOX fuel in present reactors, and make it again when needed? (bearing in mind the cost of storage)

What would be the economic cost of ensuring that fuel fabrication and utilization facilities have a greater capacity than fuel reprocessing so that plutonium can be burned up as soon as it is separated?

OTHER FUEL CYCLES

It seems that everyone who begins to study the future of nuclear power, recognizes very soon the advantages of a thorium fuel cycle *if it can be made to work.* A thorium fuel cycle does not produce plutonium, and although it produces uranium 233 from which one can also make a bomb, the uranium 233 can be quickly diluted with natural uranium for use in light water reactors. However, as soon as it is reused in reactors, plutonium is produced from the natural uranium. Twenty-five years ago, when making a breeder reactor was a national priority in many countries, the projected expansion in the need for nuclear electricity, and a smaller expectation of world uranium reserves, led to a demand for high breeding ratios so that the expansion could be rapid. A breeding ratio less than 1.4 was considered too low. This ruled out the thorium cycle. Since rapid expansion of a breeder reactor program seems not (yet) to be necessary, it is worth asking once again:

Can the thorium cycle be used instead of, or in addition to, the plutonium cycle? Does it possess the advantages that many people claim?

ACCELERATOR DRIVEN SUB-CRITICAL ASSEMBLIES

Several authors have begun consideration of sub-critical assemblies driven by spallation neutrons from a particle accelerator (Petrov 1992a, 1992b; Bowman et al., 1992; Carminate et al., 1993) Amplifications by the assembly of 50 are suggested. These ideas are not new, dating back at least to Lewis (1975). At first sight these subcritical assemblies will only solve a problem not considered here - that the public fear of nuclear reactor accidents may be related to criticality. All the fuel cycle problems would seem to be the same. However, the authors propose in addition to using a subcritical assembly to use a thorium cycle. Although not clearly stated, the design turns the safety advantage into a better assembly from the breeding point of view. Thus we should be asking:

Does the extra flexibility obtained by avoiding criticality enable us to use a more proliferation resistant cycle?

THE DANGERS OF SODIUM

The famous American physicist R.W. Wood on walking to work in Baltimore, would spit into a water pool and flick a speck of sodium metal into it at the same time to impress the small children nearby. By such means we develop a prejudice that sodium is bad. It is hard to counter this with the (correct) statement that water is far more corrosive to most materials in nature; it is mainly in corroding flesh that sodium is worse. A small sodium fire in the BN 600 reactor in Beloyarsk was headline news in the USA. Some people envisage 100 tons of sodium in a pool reactor burning out of control. The public concern with the accident at Monju, Japan, in 1996 illustrates that although sodium burning may be the smallest of worries of professionals it is a source of public concern that demands a response. Hopefully the planned IAEA meeting on sodium fires that is planned to be held in Japan will help to improve public understanding.

How does one convince firstly oneself, secondly scientific colleagues and lastly the public that the sodium danger is under control?

THE PREVIOUS ACCIDENTS

It has been said that anyone who does not learn from history is condemned to repeat it. Many people still remember the accident in the FERMI-I reactor at Laguna Beach Michigan and the famous (or infamous), but careless, statement "We almost Lost Detroit" (we were far from losing Detroit). The problems with SuperPhoenix at Creys-Malville, where the reactivity dropped suddenly without apparent reason, is a cause of concern. It seems important to be very clear what the reasons for these incidents were, and why they will not occur in any design under consideration. Therefore an important need is to:

Has anyone carefully Recorded, Understood and Explained the past history of breeder reactor technology both of accidents, and of failures and successes? If so, where is it? If not, why not?

THE EVILS OF PLUTONIUM

Plutonium is regarded by some of the US public as the embodiment of evil, and the breeder reactor has become a symbol for some public concern about nuclear power. Therefore it is important to understand this concern and how it might be allayed in the future. Plutonium has been called "the most toxic substance known to man" and while this is an untrue statement plutonium is toxic in small amounts. While it can, and should, be kept out of the environment, I note that a ton of plutonium has been evaporated into atmosphere (world wide) from bomb tests and there has been no obvious effect on health. A report written for the ANS study (Clarke et al., 1995) discusses the data in some detail.

Is the Clarke Report on plutonium toxicity widely accepted and if not, why not?

Some MAYAK workers had over 1 µCurie of plutonium in their skeletons - over 10,000 the normal amount. Although adverse effects on health were seen, they were comparable to cigarette smoking.

Can the MAYAK experience be used to help in public understanding?

NUCLEAR PROLIFERATION

It is likely that public perception will still depend on our ability to control nuclear weapons proliferation among countries. Those that espouse nuclear power and in particular a breeder reactor program which closes the fuel cycle cannot afford to ignore this issue. This is of course a major concern of IAEA and is discussed elsewhere. But I summarize what I believe to be some important points (Wilson, 1975). There have been both successes and failures. 50 years ago, when as a graduate student I first discussed this matter with those who had made the first bombs, we knew that any industrialized country could make bombs without outside assistance within a few years (I suggest one year now), and a third world country within about 10 years. The only thing that any technical non proliferation procedure can do is to slow this down a little and give the world warning that a country is proceeding in a certain way. U.S. politicians have hard time understanding these limits of technology and seem to put absolute reliance on export controls. But we must not forget that a delay of even a few years can give opportunity for diplomacy to act.

We also believed that about 100 countries would have built a bomb within 25 years (1970) but it never occurred to us that any country would have more than 10 or 20. We were wrong and too pessimistic. We did not allow for the realization by most countries that have the capability of making a bomb within a few years that an atomic bomb will not add to their security, or even to their prestige. On the latter count it is clear that the USA and USSR were both crazy. At the time of the Cuban missile crisis the USA had a few more than 100 bombs with delivery system, and the USSR a few less than 100. I was scared stiff. The increase to 30,000 bombs and 6,000 delivery systems did not scare me more. Even with the policy of mutual deterrence it is far too many. This suggests that the idea of insisting on equality is unnecessary. Any country which reduces unilaterally to 100 is doing the right thing by its own people. I am encourage that Russia and the United States have reduced the number of weapons, but we have a long way to go.

On the first count I suggest that it is instructive to understand the reasons why a country decided to make nuclear weapons. Even more important why a country decided NOT to and whether we can reinforce these reasons. A major reason why a country decided to make them was prestige. England and France did so to be taken seriously by the USA. In this the USA was at fault, and continued to be at fault for some time, in only taking seriously these countries after they had a bomb. South Africa has dismantled its few weapons, not wanting them to be controlled by the ANC. Brazil and Argentine, stimulated I am glad to say by the Physical Societies of both countries, realized that a bomb program no longer gave prestige and have abandoned their programs. It is vital for scientists to go out and talk to the potential proliferators, understand their concerns and try to address them. Alas few US scientists and diplomats do this, but instead lecture to them from Washington or academia. Iyengar (1995) noted that "as long as nuclear non-proliferation initiatives restrict their attention to the spread of nuclear materials and 'know-how' from the 'haves' to the 'have nots', without taking into account the needs, fears and capabilities of the non-nuclear states they are doomed to failure." We must consider the developing countries. I do not think the rest of the world will persuade India away from its position that NPT is a discriminatory (colonialist) treaty.

Can more imaginative, quiet, talks with India and Pakistani leaders persuade them to come to a non-nuclear agreement on their own which would satisfy the rest of the world?

I would like to see answers to these questions in various (coupled) ways. A detailed technical paper on each one; a technical survey (with reference to the detailed papers) suitable for a weekend read by any physical scientist, and a mere elementary summary for high school students. All on the World Wide Web with links to everyones' papers. Nothing less will do.

I end this part of the paper with a plea similar to that made by Andrei Dmitreyvich Sakharov to the German "Greens" at the "Congress on a Nuclear Free World" in Moscow in 1987. Rather than spend all their energies opposing nuclear electric power, Andrei challenged them to spend their energies making it safer and more secure. That seems to me to be the challenge for a forward looking scientist or politician.

THE ANTI-NUCLEAR UNITED STATES

Now I change gears and discuss briefly the U.S. Nuclear program. Even without a closed fuel cycle or completed fuel chain, nuclear power is in trouble. Many people have commented at this meeting and elsewhere (Wilson, 1990) that the inflation adjusted cost of nuclear electricity has more than doubled since 1973; both in construction and operation. This leads to the questions:

Why has the cost of nuclear power doubled in 20 years? and Why is the nuclear "industry" so complacent about the cost increase?

A large part is usually attributed to excessive regulation. The Towers-Perrin report of 1995 described some of this. I claim that the regulatory agencies are not even following their own guidelines. In 1960, at the end of a public hearing on rule making RM-30-2, the first Nuclear Regulatory Commission proposed a meaning for As Low As Practicable (ALARA). Releases of radioactive material should be reduced if that can be done for $1,000 per Man (now Person)- Rem. With a slope to the dose response curve for radiation induced cancer of 5000 Man-Rems per cancer (and a pessimistic assumption of low dose linearity) this came to $2,000,000 per "statistical life" saved. But in a study of costs Tengs et al., (1995) estimate that many NRC regulations mandate over a billion dollars per statistical life, 500 times the 1976 proposal. This leads to the question:

Why has no one objected to the strong regulations and taken NRC to court?

In the 1980s, after many hearings and discussions by the Advisory Committee for Reactor Safeguards (ACRS) the NRC promulgated a set of "Safety Goals" Among them was that core melt frequency should be kept to below one in 10,000 years per reactor. In March 1996 Northeast Utilities (NU) ran afoul of NRC. After a whistle blower had brought this to public attention, it transpired that various fuel changing procedures were being done which were not in accordance with specifications. 4 large nuclear power plants have been shut down for over 9 months at a cost for replacement fuel of well over a billion dollars with no end in sight. I have been trying to find out the substantive technical features. The NRC was slow (2 months) to respond to my questions, and then I was only told about procedural issues. Neither NRC nor anyone else has contested my rough estimate that the problems might increase core melt frequency (average value of a distribution) from 4 in 100,000 to 5 in 100,000, still well within the declared safety goals. The increase in risk for someone at the site boundary (maybe 10,000 people) might be 1 in 10,000,000, again within the safety goal and leading to a loss of "statistical lives" of one in 1000 years. This gives a cost of $800 BILLION dollars per statistical life and rising. This calculation ignores the cost in lives. Assuming that half the replacement fuel is from coal, I estimate using data in Wilson and Spengler (1996) that this NRC action has already cost 300 lives.

This procedure seems somewhat like a bus company whose drivers have been driving buses (with the knowledge and tacit approval of local police) that would not pass safety inspection. Then the Department of Transportation shuts down the whole bus system for over 9 months with a major increase in private care accidents. If that happened in New York City there would be an outcry.

Why is there no outcry about NRC safety regulatory procedures?

In the split up of the former AEC the Department of Energy (successor to ERDA) was supposed to promote nuclear power and NRC to regulate it, leading to a balance. The Carter administration changed that somewhat. The Clinton administration has changed it a lot.

The Assistant Secretary for Nuclear Energy is now merely Director of the Office of Nuclear Energy. The long term support of future reactor technology (such as the IFR) has been cut back and almost abandoned.

The administration has acted to stop a low level nuclear waste site in California which had been approved by all relevant organs of the state. This, by the way, is not the much advertised doctrine of Not In My Back Yard (NIMBY) but it is a more insidious problem not in anyone's back yard. Thus I ask the question:

Which is the most effective anti-nuclear organization in the USA?

Ralph Nader?
Friends of the Earth?
Union of Concerned Scientists?
Nuclear Regulatory Commission?
Department of Energy?
The White House?
Ourselves?

There is much talk about the effect of deregulation. But no one talks about deregulating nuclear safety.

Why not abolish the Nuclear Regulatory Commission and leave safety up to tort law and the insurance markets?

This works for oil companies. EXXON is most chastened by a multi-billion dollar judgment.

WHAT CAN WE DO?

There are few people actively pushing nuclear energy right now. In the early 1980s, a group from the Kennedy School of Government pointed out the "Director's Dilemma" explaining why no electricity company director would propose a nuclear power plant under the conditions of the time (which have got worse). Only a few lonely academics seemed to be in favor. Many of us deserted the Democratic party in the 1988 presidential election because we could not abide Michael Dukakis' anti-nuclear actions.

Now even the academics have lost hope. I have not heard the subject even mentioned in the recent elections. I cannot get anyone in the MIT Nuclear Engineering Department concerned about the fate of Northeast Utilities. It may only be the older dreamers like those of us here who have any interest left. I suggest a number of possible steps:

Suing the NRC to make them abide by their own guidelines (a public interest law foundation might help)

Asking each and every candidate for any political office his views on the subject and making the issue a deciding one in any and all elections.

Trying to ensure that our disease is not caught by other countries. This one might do by actively explaining the situation to other countries, such as France, Japan and Russia. In this connection the letter, just made public, by Professor Glenn Seaborg to the former French President, Valery Gisgard D'Estaing is of great importance.

Maybe meetings such as this might have a press conference and present a resolution on the subject.

I was educated in Oxford University, known as the home of lost causes. May be this is just another lost cause.

REFERENCES

Bowman, C.D., Arthur, E.D., Lisowski, P.W., et al. 1992, Nuclear energy generation and waste transmutation using an accelerator driven intense thermal neutron source, *Nucl. Inst. and Methods* 320:336.

Carminate, F., Klapisch, R., Revol, J.P., Roche, Ch., and Rubbia, C., 1993, An energy amplifier for cleaner and inexhaustible nuclear energy production driven by a particle beam accelerator, Report AT/93-47, CERN, Geneva.

Clarke, R.H., Nenot, J.C., Voeltz, G., Dunster, J., and Smith, H., 1995, in: Seaborg et al.; also Health Physics.

Cottrell, Sir Alan, 1981, in: *How Safe is Nuclear Energy*, Heinemann, London, UK and Exeter, NH.

Flowers, B., et al., 1975, Nuclear power, Report of the Royal Committee on the Environment, Her Majesty's Stationary Office, London.

Iyengar, P.K., 1995, Non-proliferation and advances in nuclear science, *Current Science* 68:252.

Kapitza, P.L., 1970, *J. Exp. Theor. Phys.* 30(6): 169.

Keeny, S.M., et al., 1977, Nuclear power: issues and choices", (Ford Foundation Study), Ballinger Press, Cambridge, MA.

Lewis, B., 1975, Proposal to the Canadian Government by Atomic Energy of Canada Ltd. (AECL).

Panofsky, W.K.H., et al., 1994, *Management and Disposition of Excess Weapons Plutonium*", National Academy Press, Washington, DC.

Petrov, Yu. V., 1991, A new concept of nuclear fission reactor safety, in: *Proc. 12th International Seminar on Planetary Emergencies*, World Scientific, Singapore.

Petrov, Yu. V., A new approach to fission reactor safety, *Nuclear Safety* 33:1.

Sakharov, A.D., 1978, Nuclear power and the freedom of the west" *Bull. Atom. Sci.* 34:12.

Seaborg, G.T., et al., 1995, Protection and management of plutonium, *American Nuclear Society Special Panel Report*, La Grange Park, IL.

Shprentz, D.S., Bryner, G.C., and Shprentz, J.S, 1996, Breath taking: premature mortality due to air pollution in 239 American cities, Natural Resources Defense Council (NRDC) New York, N.Y.

Singer, F., Revelle, R., and Starr, C., 1991, What to do about greenhouse warming, *Cosmos* 1:28.

Tengs, T., et al., 1995, Five hundred life saving interventions and their cost effectiveness, *Risk Anal.* 15:369.

Wilson, R., 1977, How to have nuclear power without nuclear weapons" *Bull. Atom. Sci.,* 33:39

Wilson, R., and Spengler, J., eds., 1996, *Particles in Our Air: Concentrations and Health Effects*, Harvard University Press, Cambridge, MA.

CHAPTER II

RELEVANCE OF INTERNATIONAL CONSENSUS POLICIES
ON ALTERNATIVE NATIONAL ENERGY STRATEGIES
AS RELATED TO THE REGIONAL NEEDS,
ENVIRONMENTAL IMPACT, AND GEOPOLITICAL TRENDS

ENERGY TECHNOLOGY PROGRESS FOR

SUSTAINABLE DEVELOPMENT

Dan E. Arvizu[1] and Thomas E. Drennen[2]

[1]Director, Advanced Energy Technology and Policy Center
[2]Senior Member of Technical Staff,
    Energy Policy and Planning Department
Sandia National Laboratories
Albuquerque, New Mexico 87185

ABSTRACT

Energy security is a fundamental part of a country's national security. Access to affordable, environmentally sustainable energy is a stabilizing force and is in the world community's best interest. The current global energy situation however is not sustainable and has many complicating factors. The primary goal for government energy policy should be to provide stability and predictability to the market. This paper differentiates between short-term and long-term issues and argues that although the options for addressing the short-term issues are limited, there is an opportunity to alter the course of long-term energy stability and predictability through research and technology development. While reliance on foreign oil in the short term can be consistent with short-term energy security goals, there are sufficient long-term issues associated with fossil fuel use, in particular, as to require a long-term role for the federal government in funding research. The longer term issues fall into three categories. First, oil resources are finite and there is increasing world dependence on a limited number of suppliers. Second, the world demographics are changing dramatically and the emerging industrialized nations will have greater supply needs. Third, increasing attention to the environmental impacts of energy production and use will limit supply options. In addition to this global view, some of the changes occurring in the U.S. domestic energy picture have implications that will encourage energy efficiency and new technology development. The paper concludes that technological innovation has provided great benefit in the past and can continue to do so in the future if it is both channeled toward a sustainable energy future and if it is committed to, and invested in, as a deliberate long-term policy option.

INTRODUCTION

Energy security is a vital component of U.S. national security policy. Achieving energy security requires clear differentiation between short- and long-term issues and the definition of appropriate roles for government involvement in each type of issue. For shorter term issues, assuring energy security requires proactive policies that allow the market to work. In the longer term, it requires government investment in basic R&D aimed at diversifying supply options and increasing conversion, delivery, and end-use efficiency to minimize disruptions and price volatility associated with future long-term shortages.

Technology for Global Economic and Environmental Survival and Prosperity
Edited by Kursunoglu *et al.*, Plenum Press, New York, 1997

85

This paper summarizes key issues affecting U.S. energy security, suggests appropriate government responses to both short- and long-term issues, and discusses the role technology can play in making an inevitable transition from non-renewable energy sources such as oil to alternatives. As an example of how energy R&D could lead to an energy secure future, consider the following scenario. Worldwide oil production doubles over the next 20 years due largely to increased demands of industrializing countries such as China and other non-OECD[1] Asia countries. After 20 years, worldwide oil production peaks and prices increase as supplies tighten. Meanwhile, in the utility sector, deregulation initially leads to reduced private sector funding of technology R&D as utilities slash costs to remain competitive. In subsequent years, however, deregulation leads to increased efficiency in the overall system and leads to widespread adoption of distributed power options, small generating and storage units located at strategic points along the transmission and distribution systems. Federal government and private investments in renewable energy technologies, fuel cells, and storage systems result in significant efficiency improvements that lower the overall cost and usefulness of these technologies. Transition technologies, such as fuel cells operating originally on natural gas or other hydrocarbons, achieve widespread acceptance as distributed power options. Likewise, in the transportation sector, concerns about the long-term availability of oil and environmental awareness lead to the widespread adoption of ultra low, or zero emission vehicles standards and the subsequent mass production of hybrid electric vehicles that utilize flywheels, advanced batteries, and fuel cells powered directly with hydrogen from nuclear or renewable energy sources.

This hypothetical scenario paints a future in which active involvement by the federal government in funding technology R&D reduces overall vulnerability to supply disruptions and enhances energy security. It is easy to imagine an alternative scenario as well, a scenario lacking the federal commitment to R&D. In this scenario, deregulation proceeds as above, but without federal funding for R&D, the price of alternative systems remains high. As price shocks associated with depleting reserves begin, the federal government scrambles in the face of public outcry to "do something." However, years of neglect and decay of the research infrastructure mean any response by the government will take years to have an effect. The uncertainty about supplies sends energy prices soaring in the commodity markets, leading to recession and severe economic losses for not just the U.S., but for the whole world.

Although this paper focuses largely on U.S. energy security interests, its message has clear implications for other countries as well. In terms of sustainability, it lays out a realistic plan for making the transition from a fossil-fuel based economy to more permanent, diversified energy supply portfolio for the future.

JUSTIFICATION FOR GOVERNMENTAL FUNDING FOR TECHNOLOGY RESEARCH & DEVELOPMENT

Defining Energy Security

A 1992 Department of Energy panel (DOE, 1992) on national security concluded that the U.S. will have achieved national security when:

> "...we are able to anticipate the economic, military, political, and sociocultural challenges of a diverse, interactive, changing, and unstable world order and to develop effective and appropriate initiatives and responses that benefit current and future US interests and the interests of our alliances."

Energy security is not specifically mentioned in this definition of national security; however, there is sufficient evidence that energy supply or price disruptions can have sufficient economic, military, and political implications to be considered a vital part of our national security. Clearly, there is a role for government involvement in assuring U.S. energy security.

---

[1] OECD is the Organisation for Economic Co-operation and Development and includes: Australia, Austria, Belgium, Canada, the Czech Republic, Denmark, Finland, France, Germany, Greece, Iceland, Ireland, Italy, Japan, Luxembourg, Mexico, New Zealand, the Netherlands, Norway, Portugal, Spain, Sweden, Switzerland, Turkey, the United Kingdom, and the United States.

Critics challenge this view, noting that the federal government has spent in excess of 100 billion dollars since 1973 in order to increase our energy security and that these outlays "cannot be said to have measurably altered the energy security of the nation," (Stagliano, 1995). However, this view confuses energy security with energy independence. In a world where fuels are traded in the commodity markets, reliance on foreign suppliers may be justified. As former Congressman and chair of the Subcommittee on Energy and Power of the House Energy and Commerce Committee, Phil Sharp (1995) points out, by relying on the basic market forces, "we have cheaper energy, we have greater supply, and we therefore have more energy security than if we had relied upon central control decisions to allocate energy resources."

Determining the appropriate government role in enhancing energy security requires that we differentiate between the short- and long-term and recognize that the role of government is very different in these horizons. Further, evaluating success at achieving energy security requires focusing on a broader set of criteria that includes economics, energy independence, international obligations, and the relationship of the choice to sustainability (Hartley, 1995).

Differentiation Between Short- and Long-Term Issues

Reliance on fossil fuels will continue for the foreseeable future. Although world oil prices have increased over the past few months, worldwide production capacity is adequate to meet needs for at least the short term. Despite adequate supply, short-term crises, similar to those of the past two decades, can be expected to occur. Typically, short-term crises are either politically motivated or are due to temporary market disequilibrium. Politically driven crises result from one (or a few) country's decision to reduce oil exports to achieve a foreign policy objective or as a result of wars begun as political decisions. The oil-price surges of Spring 1996 are an example of market disequilibrium type crisis. They were brought about by a series of miscalculations, including winter supply requirements, the weather, and incorrect market expectations about the timing of lifting the oil sanctions against Iraq, which led to unacceptably low inventory levels for major producers and eventually led to panic buying and sharp price increases.

Short-term crises are extremely difficult to predict. The goal of any governmental action during short-term crises should be to help provide stability and predictability in the market. Above all, the primary role for the government during short-term crises should be to allow the market to work. Oil has become a commodity bought and sold on the futures markets. Unhindered, these markets provide transparency of prices and allow buyers and sellers to hedge risks associated with price volatility. Unclear governmental policies can adversely affect this process by sending false signals to the market. This does not necessitate a hands-off policy, but rather the creation of clear policies before such crises occur. One example of an appropriate government role is prior formulation of Strategic Petroleum Reserve (SPR) drawdown plans for various short-term supply interruption scenarios.

While short-term crises are either politically driven or due to temporary market disequilibrium, there are several longer term issues on the horizon that may lead to longer term crises. These issues include diminishing oil reserves and increasing concentration of remaining reserves among a few countries in a potentially politically unstable region; fundamental changes in worldwide demographics that will mean increased demand from other world regions, such as Asia; and increased global awareness resulting in increased obligations to various international environmental agreements. Each issue is discussed in detail below, although it is often difficult to isolate the impacts associated with any one factor. For example, fundamental demographic and economic changes in developing countries complicate the task of achieving meaningful international agreements to limit carbon dioxide emissions. The long-term nature of these issues and their potential impact on overall energy security justify government involvement, because these are issues not adequately dealt with in the market place.

Oil Reserves

Many analysts incorrectly assumed that the energy crisis of the 1970s heralded the onset of declining worldwide oil production and compelled an immediate transition to alternative sources. During the 1980s, many predicted oil prices could reach $100/barrel by 2000. These forecasters failed to consider that higher prices would result in increased exploration, the discovery of vast new reserves, and the more efficient extraction, production, and use of energy.

Although past estimates of oil prices have often been wrong, fossil fuels are indeed finite, non-renewable resources. And even though estimation of ultimately recoverable reserves is an inexact science, several studies suggest that oil production will peak in the next 20-40 years. In the U.S., proven reserves have declined from 28.4 billion barrels (bbbls) in 1984 to 22.4 bbbls in 1994, Figure 1. At current production rates of 8.5 million barrels (mbbls)/day, remaining domestic reserves would only last 7.2 years.[2] Without the discovery of major new oil reserves in the U.S. or a shift away from an oil-based economy, the U.S. will have to rely more and more on imports from a select group of countries. By contrast, estimates of worldwide reserves have increased over the past decade. DOE estimates total remaining reserves of approximately 1,000-1,100 billion barrels, Table 1 (DOE, 1995a), equivalent to 43-48 years at current production rates. A review of various reserve estimates by MacKenzie (1996) suggests a slightly higher range, 1,035-1,435 billion barrels.

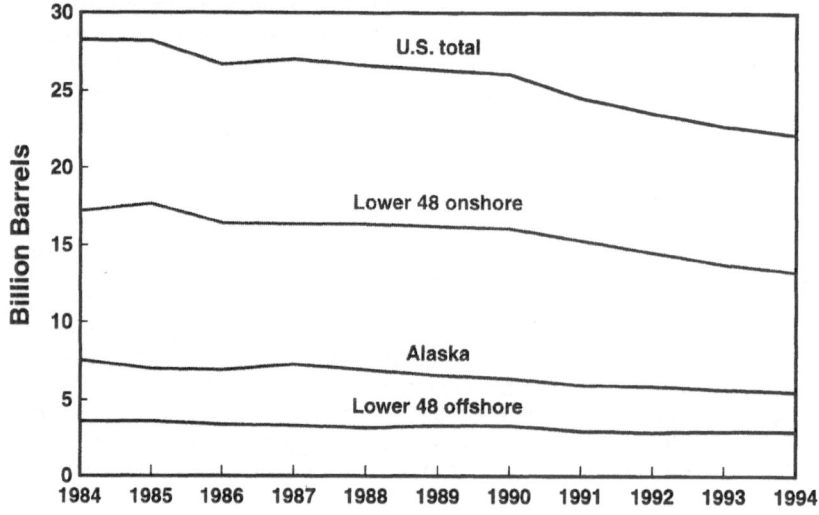

Figure 1. U.S. Crude Reserves, 1984-1994 (Source: DOE, 1995a).

Oil prices have not increased over time as forecasted in the early 1970s due to these increased reserve estimates, competition amongst suppliers, and reduced production costs. However, as is the case in the U.S., worldwide reserves will eventually peak, production efficiencies will reach a point of diminishing returns, and prices will increase. MacKenzie (1996) estimates that global production is likely to peak between 2007-2014. Even under the most optimistic assumptions about remaining reserves, MacKenzie demonstrates that production will begin to decline by approximately 2020.

---

[2] The problem with using this static measure of reserve size [referred to as the reserves/production (R/P) ratio], is that it doesn't take account of new reserves, technological change, or increased demand, thereby making it difficult to predict long-term availability.

Table 1. International Oil Reserves, Production, and Reserve/Production (R/P) Ratio

Rank	Country	Oil Reserves 1994 (mbbls)	Production 1995 (mbbls)	Max. R/P (years)
1	Saudi Arabia	161,203-262,475	3,004	87.4
2	Former Soviet Union	57,000-191,144	2,391	79.9
3	Iraq	99,427-100,000	202	N/A
4	Kuwait	96,500 - 97,675	751	130
5	Abu Dhabi	62,000 - 92,200	N/A	N/A
6	Iran	58,650 - 89,250	1,330	67.1
7	Venezuela	64,477 - 64,878	1,004	64.6
8	Mexico	49,775 - 50,776	956	53.1
9	Libya	22,800 - 36,570	507	72.1
10	China	24,000 - 30,204	1,100	27.5
11	United States	22,132 - 22,957	2,394	9.6
12	Nigeria	17,210 - 17,900	740	24.2
13	Norway	9,416 - 16,998	N/A	N/A
14	United Kingdom	4,517 - 15,492	908	17.1
15	Algeria	9,200 - 10,157	437	23.2
16	Indonesia	5,779 - 6,347	556	11.4
17	India	5,776 - 5,807	N/A	N/A
18	Canada	5,038 - 5,848	663	8.8
19	Oman	4,828 - 5,183	N/A	N/A
20	Malaysia	4,300 - 5,090	N/A	N/A
Top 20 Total for Oil		**947,917-1,063,063**		
World Total for Oil		**999,761-1,111,598**	**22,714**	**48.9**

Sources: DOE, 1996b; DOE, 1996c.

If production peaks in the next 20-25 years, this is when the U.S. and other countries need to make the transition to alternative fuel sources. Of course, the change cannot occur overnight. Therefore, responsible government policy requires an appropriate R&D program to help smooth the transition. The government should take a portfolio type approach to prioritizing research needs, selecting a broad range of projects with varying degrees of risk and potential payout. Specifically, federal R&D should not just focus on renewable technologies, but should include such areas as improved oil exploration and recovery techniques, including for oil shale; increased efficiency in energy end use; advanced nuclear design options; and cost-effective fuels or fuel substitutes for the future. The next section discusses how fundamental changes in demographics further complicate future supply questions by significantly increasing worldwide demand and altering existing consumption patterns.

Fundamental Changes in Demographics

The global energy balance will shift dramatically in the near future as developing countries pursue wide-scale industrialization. Worldwide, the EIA predicts that total energy consumption will increase 55% by 2015, from 349 Quads in 1993 to 542 Quads (DOE 1996b), Figure 2. However, that statistic only tells part of the story. The EIA predicts that 69% of that increase will occur in non-OECD or former Soviet Union or Eastern European countries. The largest increases will occur in Asian countries, with energy consumption in these increasing 150%. Despite this increased demand in developing countries, serious inequities in per capita energy consumption will remain. For example, EIA estimates assume that average per capita energy consumption in non-OECD countries, excluding Eastern Europe and the former Soviet Union, would increase from one-tenth or less of OECD country averages now to just one-fifth of these averages by 2015 (DOE, 1996b).

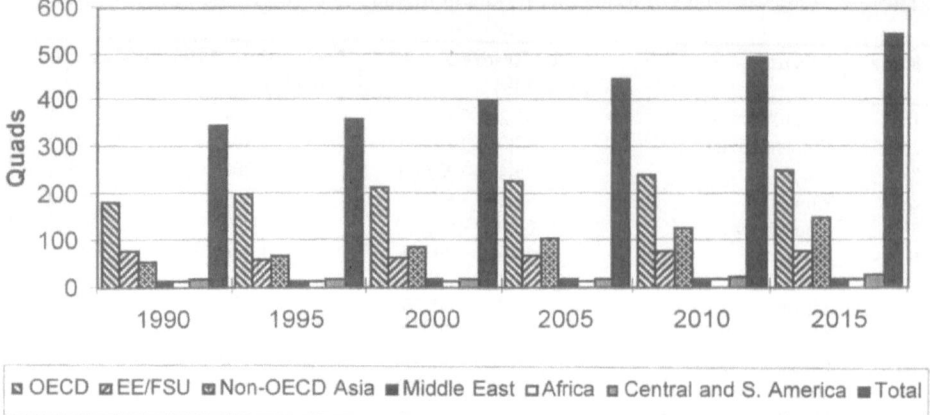

Figure 2. World Energy Consumption by Region, 1990-2015 (Source: DOE, 1996b).

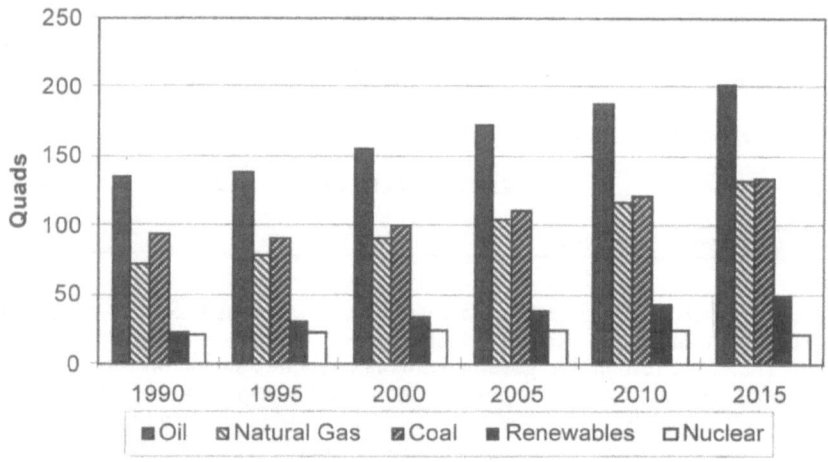

Figure 3. World Energy Consumption by Fuel Type, 1990-2015 (Source: DOE, 1996b).

Much of the increased energy demand will be for oil, Figure 3. For oil, the EIA projects that annual consumption will increase 49%, from 136 Quads in 1993 to 203 Quads in 2015. While reserves appear sufficient to meet this demand in the near term (as discussed in the previous section), increasing demand from other regions will add new stresses to the global oil supply. For example, Fesharaki (1995) estimates that oil consumption in the Asia Pacific region, which now is about 17.8 million barrels/day[3], will grow by 3.5 to 4.0 million barrels/day by the year 2000. As total demand increases, overall market share held by Middle-Eastern countries is forecasted to increase. Whereas OPEC supplied 40% of total world oil in 1993, the EIA estimates OPEC's market share could increase to 56% in 2015 (DOE, 1996a).

A specific implication of the growing world-wide demand for world oil supplies relates to production capacity. The EIA estimates that the Persian Gulf countries are currently producing at 78-84% capacity, compared to just 55% in 1985 (DOE, 1996a). The projected

---

[3] World oil consumption totaled 69.4 million barrels/day in 1995 (DOE, 1997).

increased demand in oil implies a need for increased investments in the global oil infrastructure. Otherwise, physical shortfalls due to war or other political issues could be exacerbated by this increased demand and decreased surplus production capacity, thereby possibly impacting consumers worldwide.

Compliance with International Agreements

There are many externalities associated with the production and consumption of energy. At the local level, energy production choices have a direct impact on local air quality levels. Many large cities in developing countries already familiar with deteriorating air quality conditions, will experience significant local impacts as energy production increases due to changing demographics and increasing incomes. On an international level, growing consumption of fossil fuels by all countries will increase emissions of key greenhouse gases, particularly carbon dioxide. Increased attention to international environmental problems such as climate change could, in the long term, lead countries to make the transition from fossil-based energy before actual production declines begin.

For example, at the recent second annual Conference of the Parties (COP2) to the Framework Convention on Climate Change (FCCC),[4] the U.S. announced its support for future negotiations aimed at "an agreement that sets a realistic, verifiable and binding medium-term emissions target," (Wirth, 1996). This decision to move beyond the voluntary commitments of the FCCC was motivated, according to Wirth, by the 1996 finding by the Intergovernmental Panel on Climate Change (IPCC) that for the first time "the balance of evidence suggests that there is a discernible human influence on the global climate," (IPCC, 1996).

Increasing fossil fuel consumption worldwide will make it difficult for countries to make significant progress toward reducing the threat of climate change. The atmospheric concentrations of several key greenhouse gases have increased significantly over pre-industrial (1860) levels: CO_2, 30%; methane (CH_4), 145%; and nitrous oxide (N_2O), 15% (IPCC, 1996). Achieving meaningful emission reductions will not be easy; stabilizing concentrations of CO_2 in the atmosphere at its existing level would require reducing emissions by 60% (IPCC, 1992). Based on the huge increases in fossil fuel consumption forecasted for developing countries (as discussed previously), unilateral action by industrialized nations would be largely ineffective at reducing global greenhouse gas emissions. Figure 4 illustrates the problem; despite low per capita consumption levels in developing countries, aggregate emissions from electricity production alone could grow significantly. In the next 20 years, the relative share of CO_2 emissions from the developing world, including China, could increase from 19% to 48%. Furthermore, if developing countries achieve the level of electrification expected by 2040, their relative share of CO_2 emissions would increase to over 70% of the global total from electricity production (Drennen, 1993).

The U.S., among others, has signed and ratified the FCCC. More recently, the U.S. voiced its commitment to negotiate actual binding commitments, perhaps in a manner similar to what was done for the phaseout of CFCs under the Montreal Protocol to the Vienna Convention. Although most countries still appear reluctant to commit to serious reductions in fossil fuel use, it would be difficult, if not impossible, for countries to achieve meaningful reductions in global greenhouse gas emissions without the adoption of policies that move the world away from its long-term dependence on carbon-based fuels. The eventual recognition of this issue may lead governments to enact policies that seek to persuade people, through taxes or other measures, to reduce fossil fuel consumption. Of course, increased taxes on fossil fuels in one region would decrease demand in that region, but could lead to lower oil prices worldwide, leading again to increased consumer demand, particularly in developing countries.

---

[4] Over 160 countries have signed and 120 ratified the FCCC since its negotiation in 1992. The overall goal of the FCCC is to stabilize emissions of greenhouse gases at a level that prevents "dangerous" interference with the climate system.

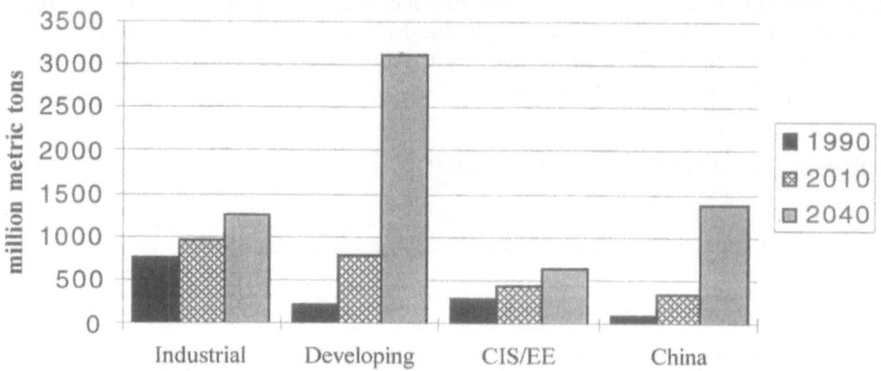

Figure 4. CO_2 Emissions from Electricity (Source: Drennen, 1993)

Summary of Role for Government in Long-Term Issues

The previous discussion summarizes the long term issues affecting global energy supply. There are no simple solutions to any of these issues. It is important to recognize the potential ramifications for U.S. energy security and to formulate a long-term policy for dealing with the fact that at some point in the not too distant future, countries will have to begin the transition away from fossil fuels and toward alternative sources. The primary goal of a country's energy policy should be to maintain price stability and predictability. As argued above, for short-term crises, this requires a clear up-front policy toward involvement in stabilizing the market, whether by use of the SPR or other measures. In the long term, achieving this goal requires, in economic terms, increasing the oil demand price elasticity so that as oil prices increase, oil demand decreases, lessening the upward pressure on prices. Accomplishing this goal would have minimal impact during short-term crises, as supply switching normally involves substantial capital costs, such as replacing furnaces, engines, or cars. However, it would have significant economic and social benefits during a sustained period of increasing prices, whether brought about by increasing scarcity, or by governmental policy (such as a commitment to reduce fossil fuel use to meet international obligations to the climate change treaty by increasing carbon taxes). Achieving this goal requires a long-term political and financial commitment to maintaining a diversified energy R&D portfolio. While past programs have attempted to do just that (alternative fuels program, etc.), success has often been linked to achieving energy independence or having a significant short-term impact. However, sound energy policy does not require that the U.S. achieve energy independence.

Energy independence should be but one of several criteria for selecting amongst competing energy sources and alternative research goals. The fact that the U.S. imports almost 50% of its oil does not necessarily imply a failed energy policy. Likewise, while reliance on domestic sources of energy could increase the overall surety of energy supply, because of the economic efficiency losses, such a policy could actually result in diminished energy security. As proposed by Hartley (1995), a more comprehensive evaluation of success at achieving energy security should include such factors as economics, energy independence, international obligations, and the relationship of the choice to sustainability.

ROLE OF TECHNOLOGY IN MAKING THE TRANSITION

The long-term issues discussed in the previous section define an approximate 20-25 year time frame for countries to begin the transition from oil to alternative sources. The existing state of affairs is not sustainable. A recent report by Shell Oil (1996) concurs; under its

"Sustained Growth" scenario of 2% annual worldwide growth in energy consumption per year, the contribution of fossil fuels (oil, natural gas, and coal) continues to increase until 2030. However, by 2065, Shell estimates that renewable energy's share could be as high as 65%. Included in Shell's estimate is a significant contribution from "surprise sources," meaning sources that are, as of now, not yet demonstrated or even yet envisioned.

Making the transition to a more sustainable future requires long-term commitments to basic and applied energy R&D. While past research has led to significant cost reductions for several technologies, Figure 5, much work remains. Our contention is that the government must play an instrumental role in spurring innovation and creativity in facilitating energy technologies aimed at this transition. Preferably, this can be done in partnership with the evolving industry where market forces drive the technology choices.

Future demands on technology innovation are great. However, the good news is that the boundaries on future advances and discoveries in energy technology are unlimited. As an example, we make the observation that to date, researchers have exploited advances in the understanding of the photoelectric effect to develop photovoltaics, the electrochemical effect to develop batteries and fuel cells, the thermionic effect to develop space nuclear power, and the basic laws governing kinetic energy to develop flywheels. In the future, we envision coupling advances in semiconductor technology to harnessing these various effects for the development of small, smart, efficient power units which could have a dramatic effect on overall end use efficiency. The point is that sufficient inefficiencies in the overall energy supply, conversion, and end use remain and technologies, both imagined and as yet undiscovered, can vastly change the future use of energy.

The next section summarizes the current status and potential of several key technologies that promise to be a part of transitional strategies in the near term.

Renewables. Research over the past 20 years has resulted in significantly improved efficiencies for renewable energy technologies. The economics (Figure 5) have improved to the point that renewables are now feasible in a broad range of niche markets, particularly remote site applications. The economics are expected to continue improving as research continues. Several of these technologies show particular promise in the developing world, where billions of people currently lack access to electricity.

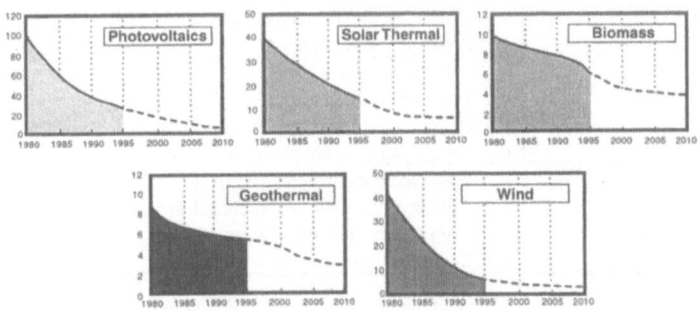

Figure 5. Estimated Costs of Renewable Energy Options in ¢/kwh (Source: DOE, 1995b)

Ongoing utility restructuring in the U.S. complicates any discussion of the future for renewable technologies here. To prepare for deregulation, utilities are cutting costs in order to lower overall production costs. In the short run, restructuring may actually increase reliance on fossil fuels as natural gas gains market share. However, in the longer term, reliance on market forces will encourage efficiency of energy supply and end use, resulting in better asset utilization. Likewise, an energy services market will emerge to better meet customer needs. This may actually accelerate the acceptance of new technologies. For example, in the long run, increased attention to improved asset utilization may lead to increased use of distributed utility options.

Distributed utilities refer to the concept of strategically locating small generating units or storage devices at points of high utilization in the transmission and distribution grid. This is important for many of the newer technologies, including renewables and fuel cells, as it creates a niche market for these new technologies that otherwise might appear uneconomical with traditional generating techniques.

While renewable energy use in the U.S. may increase as the concept of distributed utilities increases, many obstacles to its widespread deployment remain, see Table 2. Specific necessary technological improvements include advanced manufacturing techniques, integrated system designs, increased reliability, and decreased maintenance requirements. It will take a concerted and long-term commitment on the part of the government and the private sector to overcome some of these hurdles. Complicating the picture for U.S. interests is that compared to other countries, fuel costs in the U.S. are relatively low. Without a commitment in the U.S. to increased fuel taxes, or possibly externality taxes (like the proposed carbon tax), the economics of renewables may be better in other countries.

Table 2. Research/Funding Priorities for Renewable Energy Technologies

Photovoltaics	• Advanced manufacturing processes • Development of thin-film PV cells • Multibandgap, multijunction monolithic devices • Module and system reliability • International market initiatives
Solar Thermal	• High-temperature working fluids • High-temperature materials/interfaces • Thermodynamic cycles • Thermal storage materials • High-efficiency concentrators • Decreased manufacturing costs
Wind	• Wind characterization studies • Improved understanding of aerodynamics • Structural dynamics and fatigue • Improved component performance • Hybrid system design and testing • Avian risk reduction
Geothermal	• Advanced exploration methods • Improved drilling techniques • Reducing capital and operating costs • Accelerating acceptance of heat pump technology
Biomass	• Improved combustion techniques • Advanced gasification, combined-cycle technologies • High-temperature gas cleanup technologies • Short rotation intensive cropping

Storage, Peaking, and Increased Power Quality

As discussed above, utility restructuring will give rise both to markets for better asset utilization and custom power. Central to these market needs are storage/peaking/power quality systems. At present, among the most interesting options are batteries, superconducting magnetic energy storage devices (SMES), and flywheels. While most of the attention in the past has focused on batteries, SMES and flywheels may also hold great promise for the longer term. For example, existing batteries are limited to storage of about 1.5 kwh/sq. foot. It is expected that the storage capabilities of advanced batteries may be in the range of 5-10 kwh/sq. ft. Flywheels, which would store energy in the form of kinetic, rather than chemical energy, have a projected range of 10-14 kwh/sq. foot. As with the renewable technologies, significant research obstacles remain. Table 3 summarizes the current status, potential, and research needs for the storage technologies.

Table 3. Status of Storage Technologies

Technology	Status	System Size	Application Status	Technology Need	Energy Density
Batteries	Turnkey systems commercially available	kw - MW	Utility-scale applications in US: Chino & Puerto Rico	Improved cycle life and energy density Lower-cost "modular" systems	Existing Batteries: 1.5 kwh/s.f. Advanced Batteries: 5-10 kwh/s.f.
SMES	Turnkey systems commercially available in Micro-SMES	1-6 MW-sec	5 Micro-SMES units are in use today	Improved high temperature superconductors Less expensive manufacturing techniques High-powered switching device development	SMES systems are generally power systems
Flywheels	Privately funded research on components	Less than 40 kw	Complete utility-scale system demo is at least 5 years away	High-strength, lower-cost rotor materials Long-life, low-loss bearings System integration & interconnection capability	Projected: 10-14 kwh/s.f.

Fuel Cells. Fuel cells, perhaps more than other technology, could well be a technology that assists in the eventual transition away from fossil fuels. Fuel cells directly generate electricity through an electrochemical reaction between a fuel, such as hydrogen, and an oxidant. The reason that fuel cells may be a true transition technology is that in the near term, they can run on reformed natural gas (with potential to run on other petrofuels) until such time as it becomes practical to rely on clean, non-carbon based sustainable technologies, such as nuclear or renewables, to provide inexpensive hydrogen. In a fuel cell operated by natural gas the gas is first processed to create a hydrogen-rich gas that is then electrochemically reacted with an oxidant to provide power.

In addition to assisting the transition from natural gas to alternative energy sources, such as renewables, there are other benefits that make fuel cells attractive to utilities, independent power producers, and automakers. Fuel cells fit in well with the distributed utility concept: their modular design allows their placement in some applications in the transmission grid.

Present-day fuel cells operate at elevated temperatures of 200-600°C and thereby provide a good technology match when there is a cogeneration application needing electricity and heat. Finally, fuel cells also provide environmental benefits, typically, emitting just 1% of the NOx emissions associated with internal combustion engines and 10% of those associated with gas turbines.

Widespread commercialization of fuel cells requires advances in several areas, including improved manufacturing techniques to reduce system costs, new catalysts to replace expensive precious metal catalysts, improved membranes, and technologies for producing low-cost hydrogen from non-fossil fuel sources.

Other technologies. In addition to renewables and fuel cells, there are many other technologies that may be important for managing the transition from today's fossil fuel dominated economy to a more sustainable economy. Some of those technologies, such as nuclear, already play a significant role in meeting our energy needs. For example, while nuclear faces many obstacles to continued growth in the U.S., nuclear's market share continues to increase worldwide. A total of 437 nuclear reactors generated over 22% of the world's electricity in 1995; an additional 85 plants are under construction, including 32 in the Far East (DOE, 1996d).

Finally, on even a longer scale, continued support of energy R&D could lead to surprises in energy technology. For example, fusion remains an attractive option for the next century and beyond. Remaining obstacles are significant but may be overcome through ongoing international collaborations such as the International Thermonuclear Experimental Reactor (ITER) program.

CONCLUSIONS

Energy security is a vital part of national security. Assuring energy security requires clear differentiation between short- and long-term issues and a definition of the appropriate role for government involvement in each time horizon. In the short term, reliance on fossil fuels will continue. Despite adequate supply, short-term crises, similar to those of the last two decades, are likely. Short-term crises are extremely difficult to predict and provide limited opportunity for government involvement. However, in the long run, reliance on fossil fuels is not sustainable. Long-term issues leading to this conclusion include diminishing oil reserves and increasing concentration of remaining reserves among a few countries in a politically unstable region, fundamental changes in worldwide demographics shifting oil demand toward the Asian countries, and increased global awareness resulting in increased obligations to various international agreements.

These long-term issues define a 20-25 year time frame for countries to begin the transition to alternative sources of energy. Making this transition to a more sustainable future requires long-term commitments to basic and applied energy R&D. Past research efforts have led to significant improvements in existing technologies, as evidenced by the large efficiency gains and improved economics of renewable energy technologies. In the future, the government must play an instrumental role in spurring innovation and creativity in energy technologies aimed at this transition. Achieving a stable long-term energy supply will place heavy demands on technology innovation. Fortunately, as evidenced by some of the technologies highlighted in this paper, the boundaries on future advances and discoveries in energy technologies is unlimited.

ACKNOWLEDGMENTS

The authors would like to thank the following individuals for their thoughtful comments and stimulating discussion on previous versions of this paper: Abbas Akhil, Arnie Baker, Wil Gauster, Dan Hartley, Dick Lynch, Tom Mancini, Bill McLean, Nestor Ortiz, and Anne Van Arsdall. In addition we thank Cathy Ehgartner for her assistance in preparation of the manuscript.

REFERENCES

Drennen, T., 1993, *Economic Development and Climate Change: Analyzing the International Response,* Ph.D. Dissertation, Cornell University, Ithaca, NY.

Fesharaki, F., 1995, Energy outlook in the newly industrialized Asian countries, in *Energy and National Security in the 21st Century* , P. Clawson, ed., National Defense University Press, Washington, DC.

Hartley, D., 1995, *Renewables: A Key Component of our Global Energy Future,* paper presented at the International Conference on Economics and Politic of Energy, Miami Beach, FL, November 27-29, 1995.

Intergovernmental Panel on Climate Change (IPCC), 1996, *Climate Change 1995: The Science of Climate Change*, Cambridge University Press, Cambridge.

Intergovernmental Panel on Climate Change (IPCC), 1992, *Climate Change 1992: The Supplemental Report to the IPCC Scientific Assessment,* Cambridge University Press, Cambridge.

MacKenzie, J., 1996, Heading off the permanent oil crisis, *Issues in Science and Technology,* (summer 1996).

Sharp, P., 1995, Do energy imports matter: the political view, in *Energy and National Security in the 21st Century,* P. Clawson, ed., National Defense University Press, Washington, DC (1995).

Shell International Limited, 1996, *The Evolution of the World's Energy Systems,* Shell International, London.

Stagliano, V., 1995, The ghost of OPEC, in *Energy and National Security in the 21st Century,* P. Clawson, ed., National Defense University Press, Washington, DC.

U.S. Department of Energy, Energy Information Administration, 1995a, *U.S. Crude Oil, Natural Gas, and Natural Gas Liquids Reserves 1994 Annual Report,* October 1995.

U.S. Department of Energy, *Office Energy Efficiency & Renewable Energy 1995*, 1995b, Office of Utility Technologies.

U.S. Department of Energy, Energy Information Administration, 1996a, *Oil Production Capacity Expansion Costs for the Persian Gulf,* DOE/EIA-TR/0606, January 1996.

U.S. Department of Energy, Energy Information Administration, 1996b, *International Energy Outlook 1996,* DOE/EIA-0484(96).

U.S. Department of Energy, Energy Information Administration, 1996c, *Annual Energy Review 1995,* DOE/EIA-0384(95).

U.S. Department of Energy, Energy Information Administration, 1996d, *World Nuclear Power and Fuel Cycle Report,* November.

U.S. Department of Energy, Energy Information Administration, 1997, International Energy Annual, DOE/EIA-0219(95).

U.S. Department of Energy, Office of Intelligence, 1992, *Identifying Vital Issues: New Intelligence for a New World,* J. Glicken and D. Engi, eds., June 1992.

U.S. Department of Energy, Secretary of Energy Advisory Board, 1995, *Energy R&D: Shaping our Nation's Future in a Competitive World, Final Report of the Task Force on Strategic Energy Research and Development.*

Wirth, T., 1996, Statement by Under Secretary for Global Affairs, at the Second Conference of the Parties to the Framework Convention on Climate Change, July 17, 1996, Geneva, Switzerland.

UNDERSTANDING ENERGY AND GREENHOUSE GAS TRENDS, THE IEA APPROACH

Jean-Marie Bourdaire

Director
Office of Long-Term Co-operation and Policy Analysis
International Energy Agency
OECD
Paris, France

INTRODUCTION

Under the general title of this session "Economics and Politics of an Optimum Energy Demand Reconciling the Needs of the World and Security" I will concentrate on four main issues: the demand for energy, the role of oil, the case of gas and the environment which will also be my conclusion.

Technology for Global Economic and Environmental Survival and Prosperity
Edited by Kursunoglu *et al.*, Plenum Press, New York, 1997

99

Evolution of the World Energy Demand: a Sectoral View
1971-1994

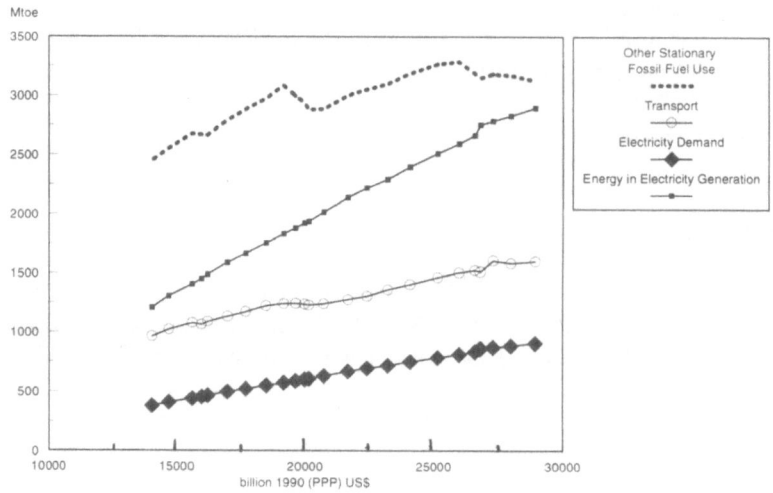

Figure 1.

First what is demand?

- Fulfilling 3 needs:
  Electricity
  Mobility
  Heat

As you can see, even with the two oil shocks and the major restructuring which has taken place since 1990 in the countries in transition (former USSR, Central and Eastern Europe), it appears that there is very little room for manoeuvre for electricity and mobility which are nearly straight lines.

There is some evidence of flexibility for the heat demand but this flexibility only occurs at the time of major economic dislocations. Clearly it is not what the world is looking for.

This confirms that the single most important parameter by far is economic growth.

Energy Demand in IEA Countries: a Sectoral View
1960-1994

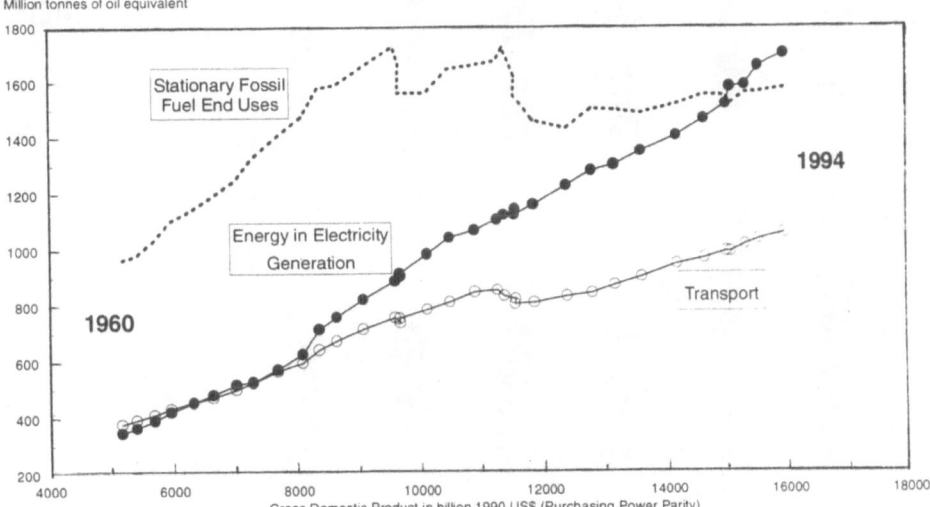

Million tonnes of oil equivalent

Stationary Fossil
Fuel End Uses

1994

Energy in Electricity
Generation

Transport

1960

Gross Domestic Product in billion 1990 US$ (Purchasing Power Parity)

Figure 2.

When focusing on OECD one finds the same drivers:

- mobility which is rising with a very small inflexion at the time of the second oil shock when, in
  North America, the CAFE standards became enforced at a time of rising gasoline prices.

- electricity which is the fastest and most rigid trend.

- heat where you can notice the sharp drop at the time of the two oil shocks and the resulting
  change of the slope of the trend which was rising as fast as electricity in the 60's and early 70's and
  is nearly flat since 1982, thanks to a combination of factors including industrial delocation,
  efficiency policies and end of the transformation of non-commercial energy into commercial
  energy.

THE IRRESISTIBLE RISE OF ASIA
Relative share of GDP of the different regions

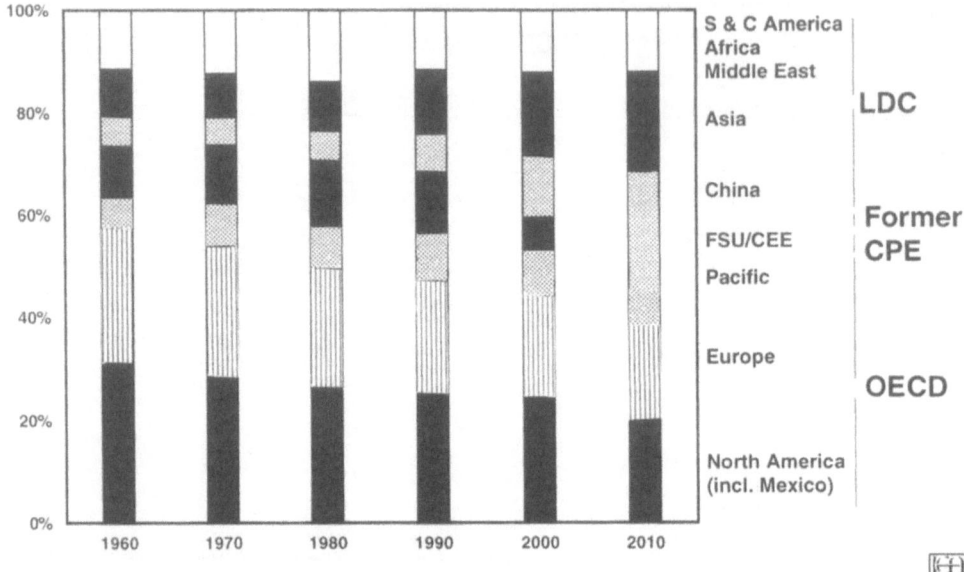

Figure 3.

Demand is the extrapolation of these trends (which we have just briefly described) according to the forecasts of economic growth.

Needless to say, this forecast is the most sensitive parameter and the one which, in the past, has tended to be overestimated. That of the IEA could also prove finally to be optimistic yet the share of growth among the different parts of the world is likely to be correct.

What does this forecast say:

- FSU/CEE economies are shrinking enormously
- OECD countries are losing ground
- Asia including China is the big winner

THE WORLD ENERGY AND OIL MECHANISM*

      ■ **Total Energy Demand (linked to world GDP)**

minus ■ **Supply of "unavoidable" energy (hydro/nuclear/coal/gas/oil outside Gulf countries)**

equals ■ **Call on Gulf OPEC**

Oil, and in particular, Gulf countries oil plays the role of swing-supplier

*Excluding FSU/CEE

Figure 4.

Let us turn now to the supply side and to the role of oil. For each basic need, mobility, electricity and heat, oil is used at the margin because its variable cost is the highest among all primary energies.

Mobility is nearly 100% oil
in electricity generation, oil is used for peaking facilities either under the boiler (heavy fuel oil) or in turbines and small diesel units (distillates)
in the heat sector, oil is used in the same way as for the power plants.

Overall, oil is the marginal fuel. However, as the supply of oil of most countries is not constrained (oil flows at the maximum production rate) and is therefore as "unavoidable" as that of hydro, nuclear, coal or gas which are cheap to run, this swing role tends to concentrate on the oil producers of the Middle East.

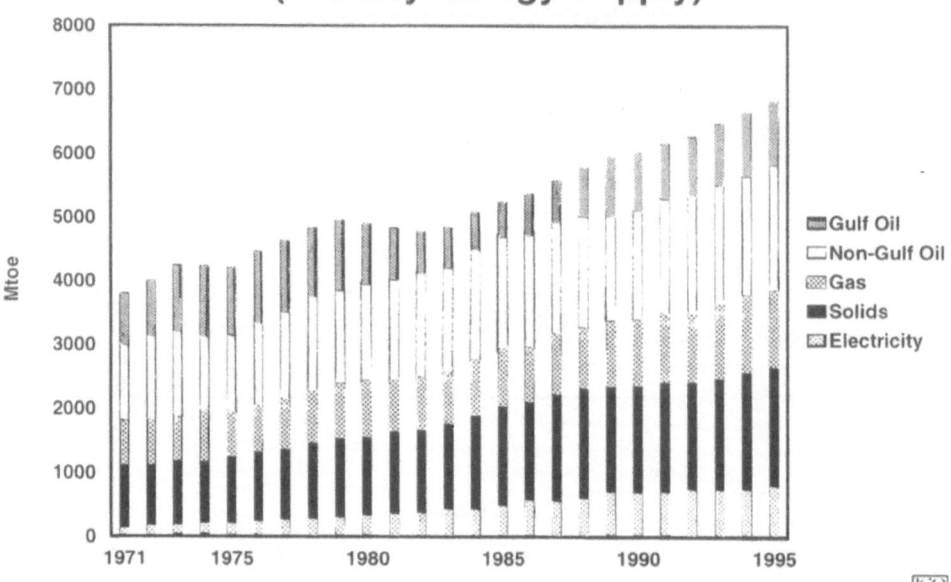

Figure 5.

This slide illustrates the previous equation and one can easily see that the Arabo-Persic Gulf oil is making up the swing supply.

OPEC: THE ACTORS

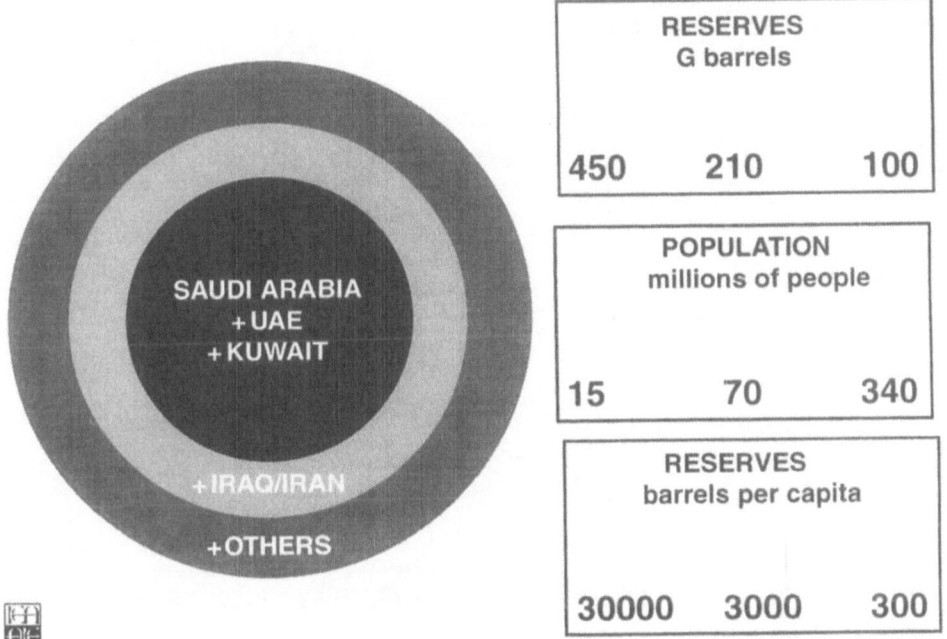

Figure 6.

Let us summarize:

- oil is the swing supply of energy
- within the oil supply, this role of swing is concentrated on the Middle East countries i.e., the two
 internal rings of this sketch.

Now, within the Middle East it appears that the inner ring, that of the Arabic Peninsula plays the greater role because of its concentration of reserves over a limited number of inhabitants.

It is even possible to push this reasoning even further to discover that, within these few countries of the Peninsula, that which is at the core of the role of swing supplies is Saudi Arabia.

OPEC PRODUCTION AND US IMPORTS

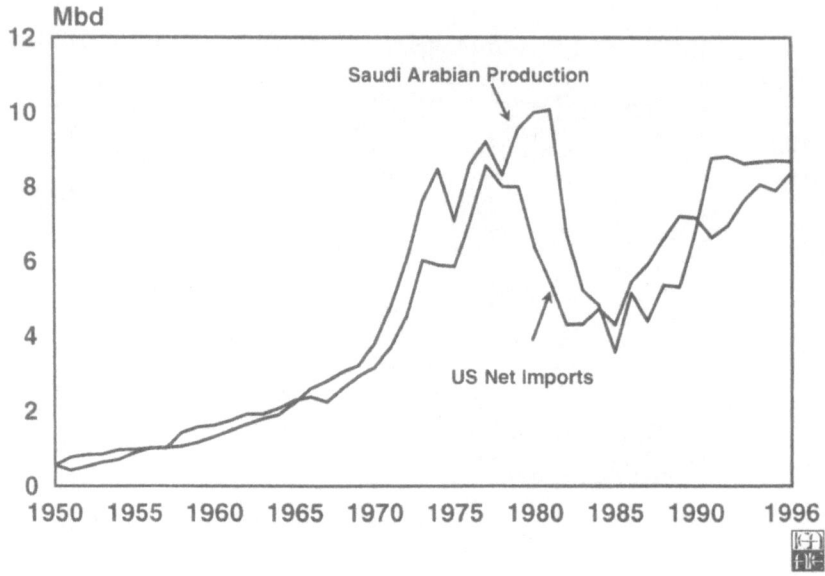

Figure 7.

That role of Saudi Arabia is generally known. What is less known is , symmetric on the demand side, that of the US.

There are many features which explain the very special role of the US on the demand side. However, I would like to focus on one factor which, in my personal view, is perhaps the most important one: whereas other importing countries, in Europe or Asia-Oceania, have increased the price of oil products to the final consumers in order to make up, implicitly or explicitly, for the security of supply, the US has chosen another way, that of military power. The cost of security of supply is not paid by the gasoline consumer, it is paid by the tax payer as a contribution to the defence budget.

WORLD HYDROCARBON RESERVES

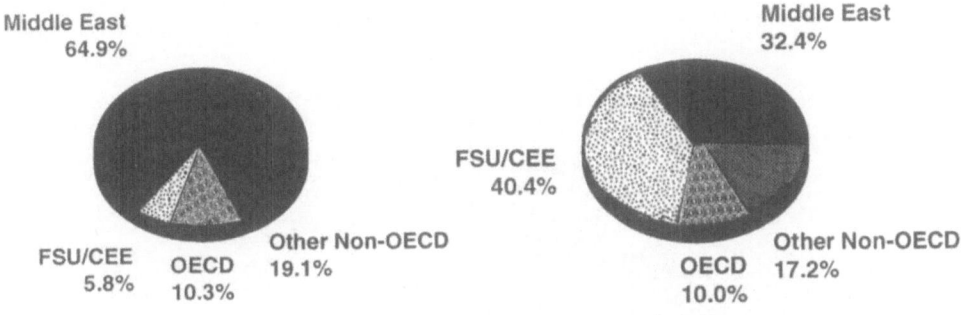

Middle East
64.9%

FSU/CEE
5.8%

OECD
10.3%

Other Non-OECD
19.1%

Oil
1 017 Gb
43 years of production

Middle East
32.4%

FSU/CEE
40.4%

OECD
10.0%

Other Non-OECD
17.2%

Gas
879 Gboe
65 years of production

Figure 8.

As a transition from oil to gas, let us have a look at the reserves and the security issues they suggest.

Oil

More than 70% i.e., Middle East and FSU are "official" figures. While the true potential is probably huge, it is not really known and could be overestimated.

One should keep in mind that OPEC figures were raised during the 80's without any geological evidence. It was just pure internal politics to increase each country's bargaining power for the quota allocation. In the same way little is known about Iraki reserves where all exploration stopped in 1959.

Gas

More than 70% is not only uncertain for methodological reasons (Middle East and FSU) it is also remote or very remote from their potential markets. That means huge transportation costs, autoconsumption and CH4 losses which all bring the true GHG content of this gas to a level which is not far from that of other fossil fuels.

At the same time, there is strong evidence of the workings of technical progress, market competition and breakthrough technologies. Geological and geographical limits tend to be pushed further away thanks to these workings. How far? We are presently investigating that issue.

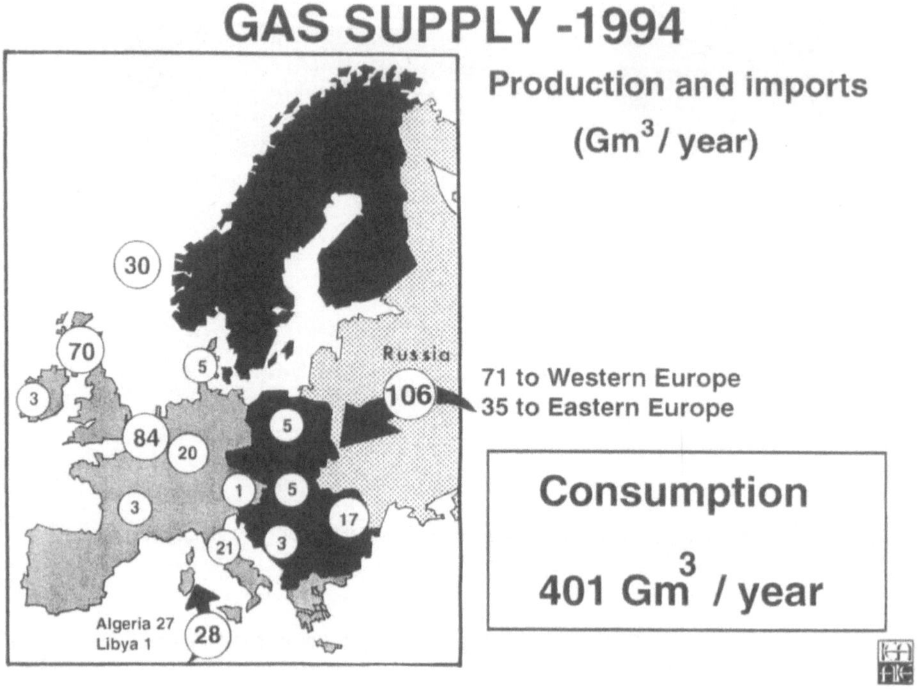

Figure 9.

Let us concentrate on Europe as an example.

In 1994 one third of the supply is imported with 106 Gm3 coming from Russia and 28 Gm3 from North Africa.

GAS SUPPLY -2010

Production and imports
(Gm^3/ year)

Consumption
628 Gm^3 / year

Figure 10.

The IEA capacity constraints case forecast a huge increase of consumption, over 50% mainly in the power sector.

Although the indigeneous production is envisaged to grow by more than 20% despite the plateauing and decline of continental Europe excluding Norway, the external dependancy will reach nearly half (as compared to one third) with Russia's share growing from 25% of the total supply to one third if Middle East imports are channelled through the Russian pipe line systems.

INCREASING ENVIRONMENTAL PROTECTION

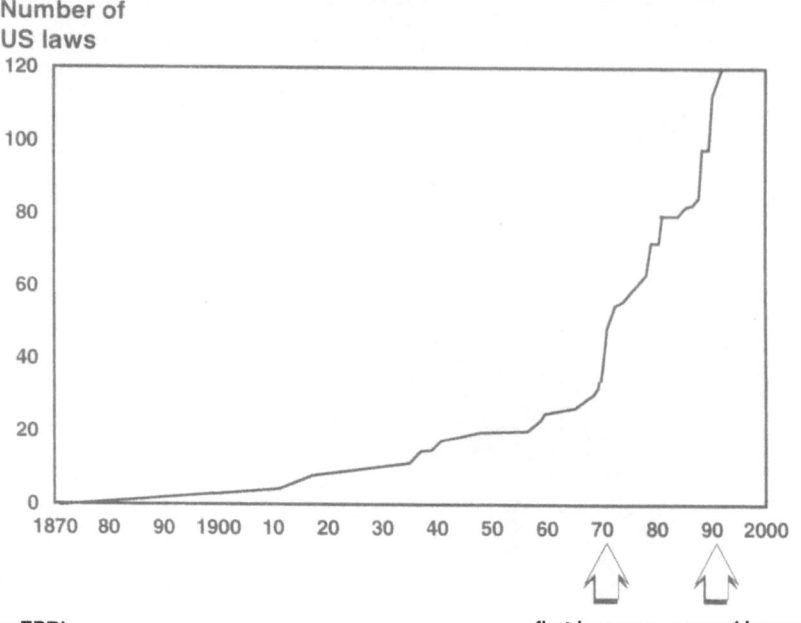

Source: EPRI   **first increase   second increase**

Figure 11.

I will not repeat my remarks and arguments about Climate Change but will rather concentrate on the US and on its nuclear industry as it is the main topic of this conference.

This slide, a courtesy from EPRI, suggests the growing importance of environmental factors. For instance, it is my personal view that the fundamentals of the first oil shock have their roots in the impact of the pollution of Santa Barbara beach in California in 1969 because it triggered all the big environmental laws of the early 70's and, as a direct consequence, the slow down of the US indigeneous energy supply.

As far as nuclear is concerned let me tell you two very simple personal remarks about the possible reasons of its US failure. Firstly, it occurred at a time of low energy prices (60's) and abundant coal supplies. Therefore it was not driven by the market and was rather an artificial move. Secondly, a lot of brakes and costs were added because of the burden of a complex and ever changing regulation framework.

That means that my plain conclusion will also be two-fold:

Firstly: there is a need for any further nuclear rise to be supported by the market. In other words such a rise is unlikely to occur unless fossil fuel prices rise either because of a supply crunch or/and because of the Climate Change policies. There is, of course, the exception of a few developed or developing nations willing not to depend too much on imported fossil fuels.

Secondly: there is a need to get the support of the national administration. Unless clear rules of the game, in terms of safety, standards and cost effective regulation, are implemented, nuclear energy will find it difficult to rise again.

Thank you for your attention.

110

THE IEA APPROACH TO CLIMATE CHANGE

Jean-Marie Bourdaire

Director
Office of Long-Term Co-operation and Policy Analysis
International Energy Agency
OECD
Paris, France

Technology for Global Economic and Environmental Survival and Prosperity
Edited by Kursunoglu *et al.*, Plenum Press, New York, 1997

Climate Change Agenda:
IEA as an "honest broker"

Policy-makers

IEA Secretariat

National private/public expert bodies

Agence Internationale de l'Energie

International Energy Agency

Current climate commitments

- 1992, Rio Summit sets a goal for OECD and economies in transition (Annex I):

    Return greenhouse gas emissions to their 1990 level by 2000

  - Current outlook: few Parties will meet their Rio commitment

  - Next step: Kyoto 1997, negotiated *protocol* towards strengthened commitments (Annex I)

The United Nations
Framework Convention on Climate Change

- A *negotiation* body with government participation from:
  - Environment Ministries
  - Foreign Affairs

- Has lacked energy sector expertise

- As a result: the 1992 Rio target is apparently unfeasible. Exceptions:
  - Germany: re-unification and economic restructuring
  - UK: 'dash for gas' in power generation

Failure to meet Rio Target: a surprise?
No, but...

- Several forecasts indicated a range above 1990 levels (including the World Energy Outlook) - Did they lack credibility?

- Perceived as *black boxes*, with energy efficiency as a central, yet opaque element

- A new approach is needed

Elaborate an energy frame for the FCCC

- Blaise Pascal: *"On ne se convainc qu'avec ses propres arguments"*

- Bring a set of insights on energy to non-energy specialists:

  – physical capital stocks constrain the timing of reductions
    - lifetime of existing infrastructures
    - lead time of new technologies

  – economic cost will be considerable if rigidities are not accounted for.

Energy, CO2 and climate:
The need for sectoral analyses

- Electricity (generation and end-use)
- Transport
- Stationary fossil-fuel combustion end-uses

Elements for analysis:

○ the dynamics of capital-stocks
  potential for savings and fuel-switching
  lessons from past energy and environment
  policies

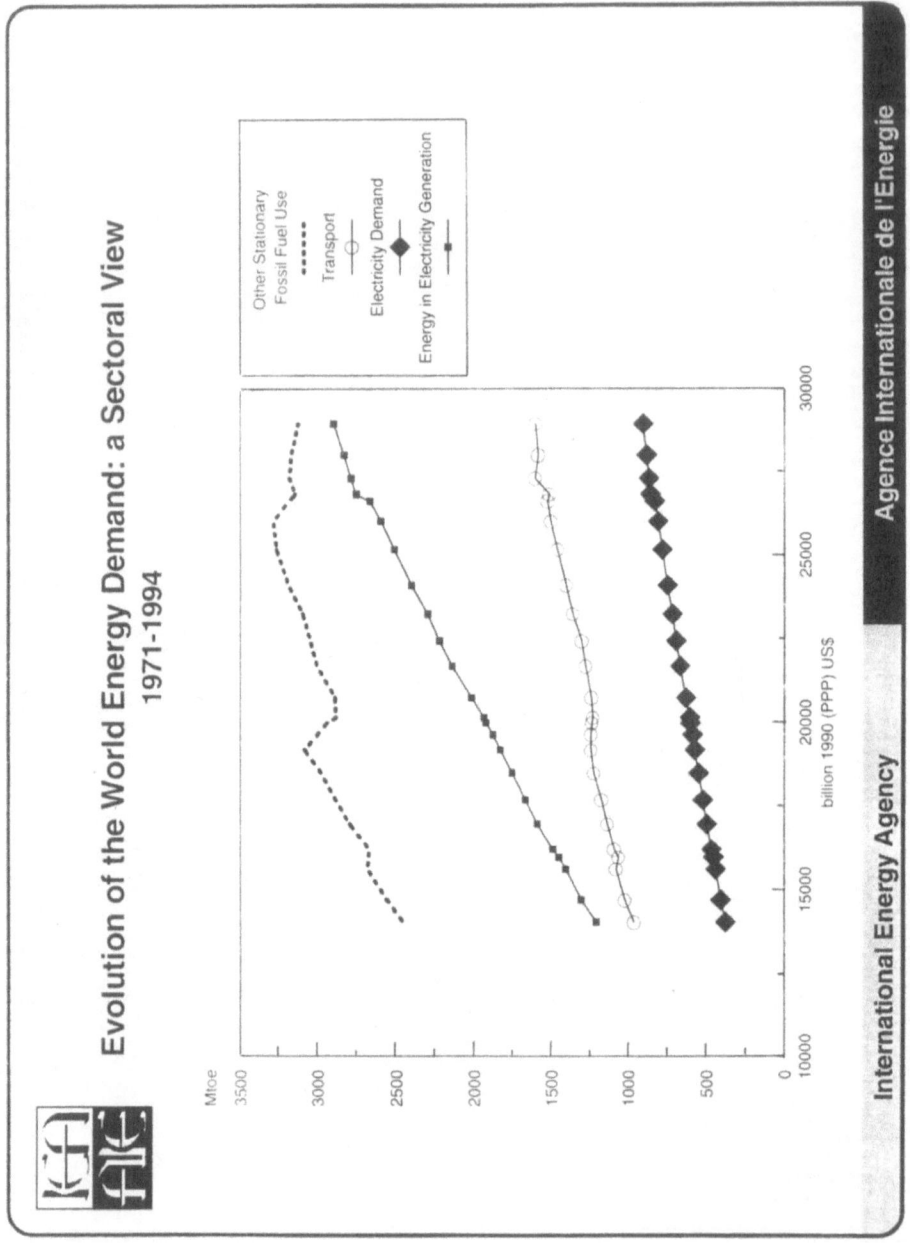

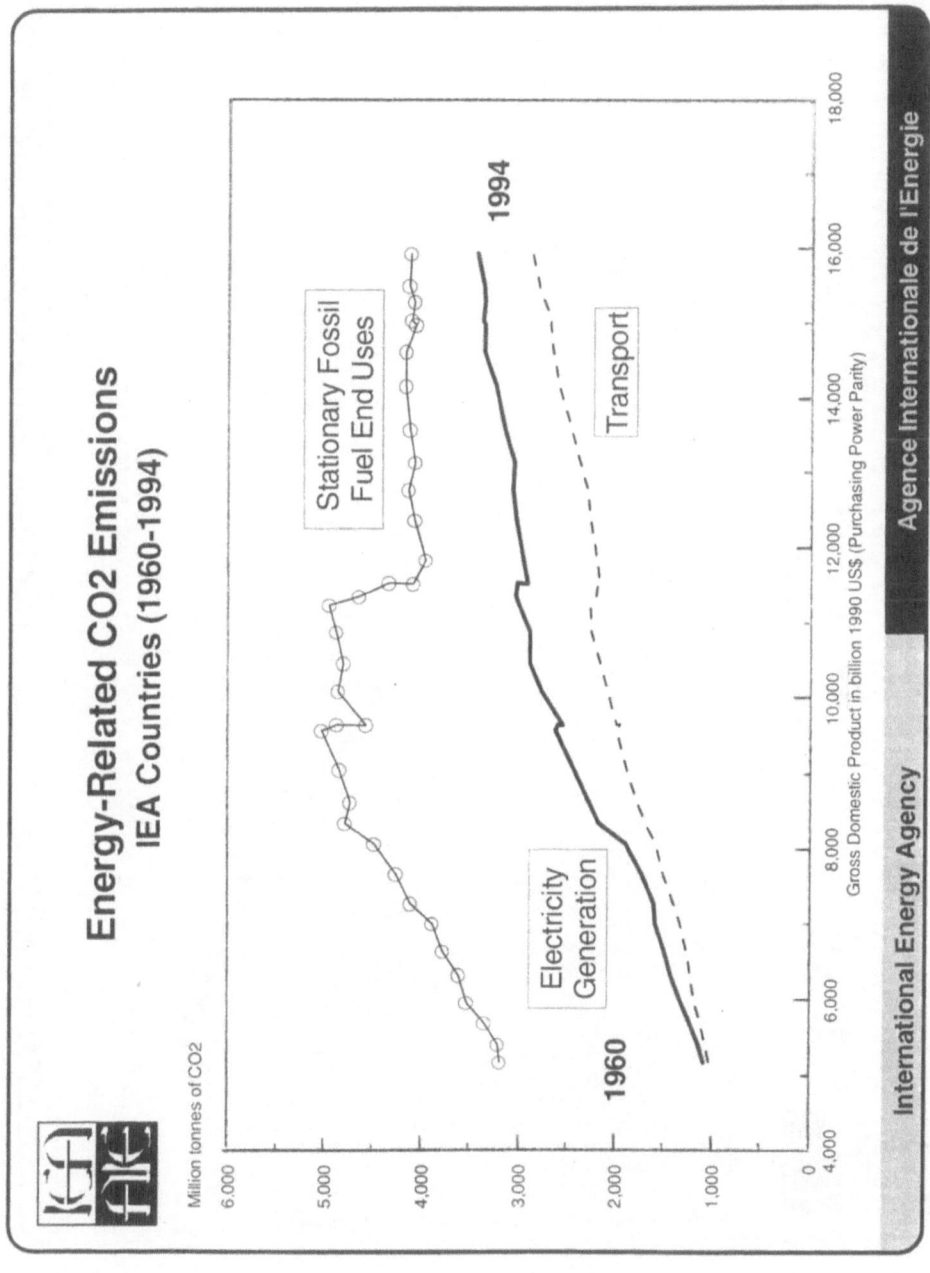

Energy-Related CO2 Emissions
IEA Countries (1960-1994)

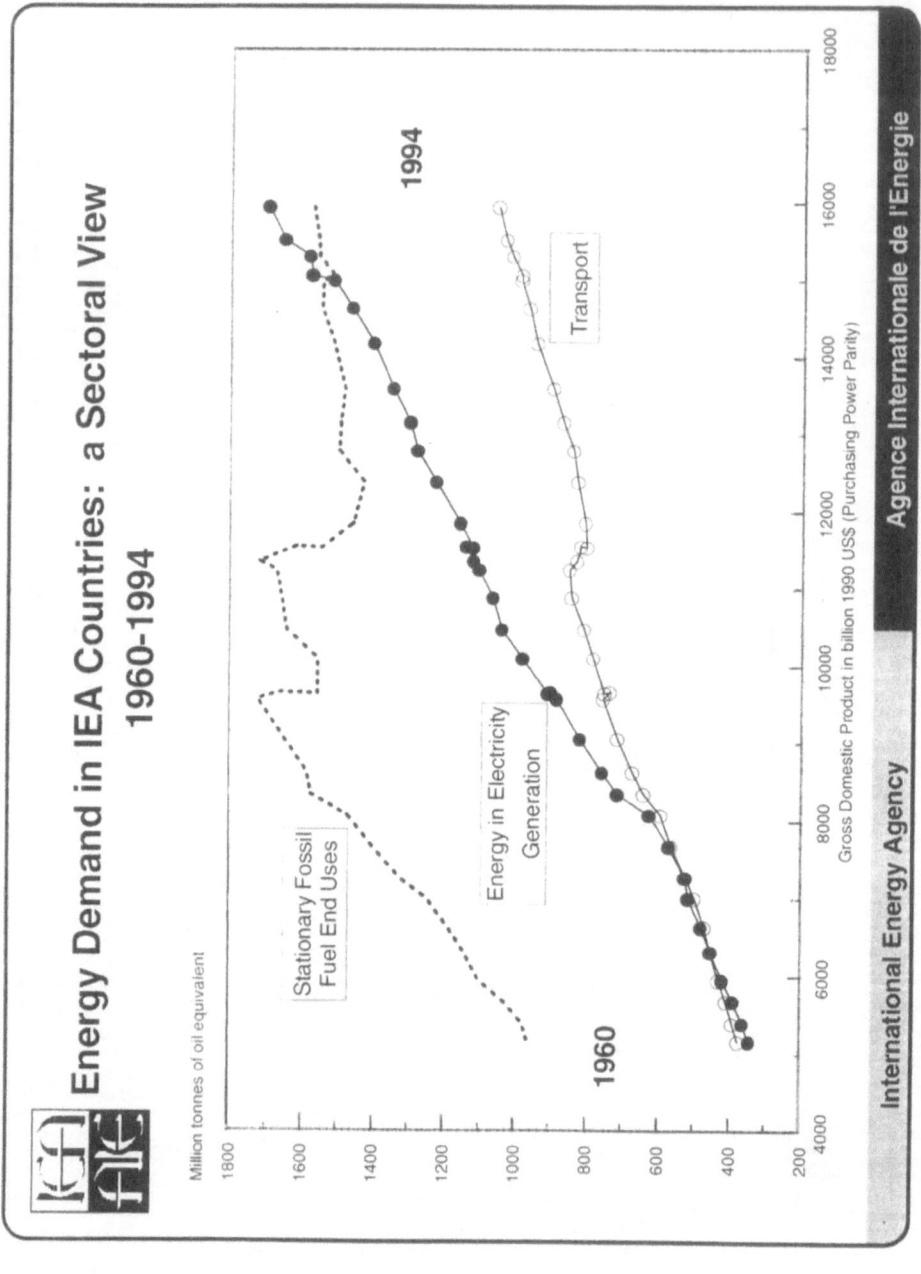

Energy Demand in IEA Countries: a Sectoral View 1960-1994

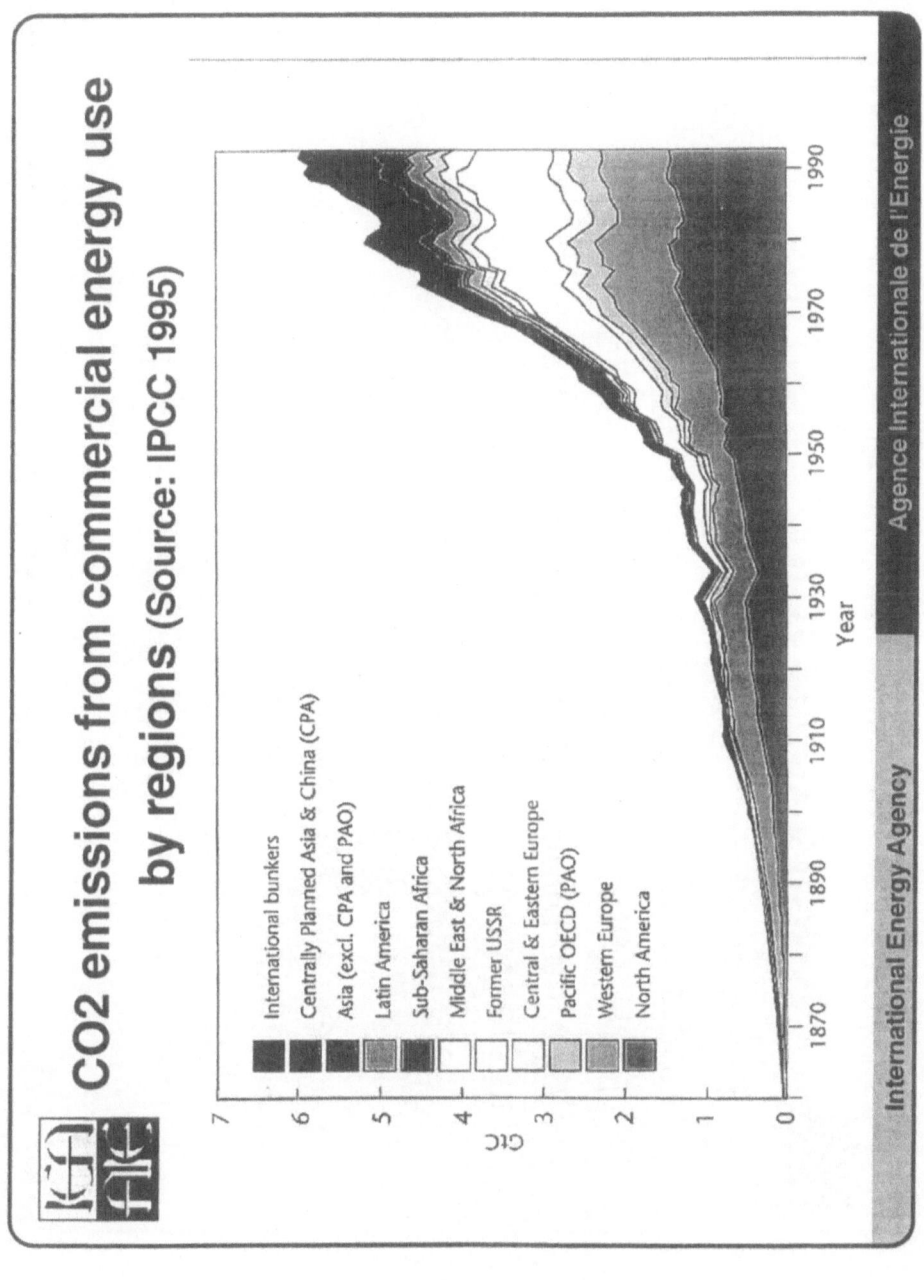

CO2 emissions from commercial energy use by regions (Source: IPCC 1995)

The IEA Approach:
towards a sectoral expertise

- From 1995 on: mobility and electricity dynamics identified as main issues

- Early 1996: Presentation of trends to the Informal Ministerial Meeting on Climate Change in Denmark

- Late 1996: call for a document for the FCCC: "The Energy Dimension of Climate Change"

- - Second "Modelling" seminar on 'no-regrets' and minimum-cost options by sector
  - Reconcile World Energy Outlook, Energy Dimension of Climate Change and No-regret

An IEA contribution:
The Energy Dimension of Climate Change

- IEA Energy Ministers are fully aware of current climate change commitments

- They expressed the need for a positive, concrete and realistic input to the FCCC process from the energy world

IEA's *'Energy Dimension of Climate Change'*

Energy Dimension of Climate Change

- OECD and non-OECD: The contribution of commercial and non-commercial energy

- Trends in the OECD: sectoral dynamics and constraints in fossil energy use

- Lessons from experience for viable and cost-effective policy responses to climate change

Several policy options for a similar carbon reduction objective

Illustration: reduction of fossil fuel use in transportation could be achieved through...

- A carbon tax on gasoline

- A 'feebate' on the car prices based on their fuel economy

- A standard on car fuel economy *à la* CAFE

The need for a sectoral approach: carbon taxes and capital cost

Over the lifetimes of these 'equipments' a $20 carbon tax is <u>roughly</u> equivalent to...

1%...

of the price of a car

25%...

of the capital cost of a coal-based power plant

20%...

of the cost of a domestic gas heater

Main Policy conclusions

- Sectoral differences should not be ignored

- Equity as a central issue in the negotiation
  - OECD/Annex I Parties and developing countries
  - *within* IEA Member Countries, given a wide range of different energy situations (starting points)

- What are the possible approaches under the Berlin Mandate:
  - Legally-binding quantified objectives (QELROs)
  - Co-ordinated policies and measures

Elements for a cost-effective and equitable protocol

- Need to recognize different energy situations in OECD/IEA countries with respect to CO2:
  - structure of economy (energy intensities)
  - fossil fuel endowments, energy pricing and regulations

- Need to go with the flow of decision-making on physical equipment:
  - avoid premature write-off of capital stock
  - incorporate less carbon-intensive processes

CHAPTER III
EDUCATION ON ENERGY AND GLOBAL ISSUES

THE IMPORTANCE OF EDUCATION
FOR NUCLEAR ENERGY

Gerald Clark

Secretary General
Uranium Institute
London

When I offered to prepare a paper on this subject I had not realised how difficult it would be. I was inspired by a feeling that compared with my own youth the present generation has not been encouraged to learn about nuclear science. The exciting facts about nuclear energy are left in obscurity, even though nuclear electricity is now a mature industry producing 17% of the world's electricity, in contrast to the situation 40 years ago when it was still largely a dream to harness the amazing power of the atom for civil purposes. Perhaps bombs are more exciting than the electricity which comes out of the wall, but most of the material progress which we have experienced in the past half century has been due to the progressively greater application of electricity.

It seemed to me that education in nuclear science was being neglected. Children learn all about micro-organisms, or the progress of medical science since the days of Edward Jenner and Pasteur. But do they learn about the experiments of Henri Becquerel and Marie Curie? Do they know why Otto Hahn found it difficult to believe that the result of the fissioning of uranium was barium. Is it then surprising that nuclear power is not as popular as we might want it to be? I have been struck, in the short time that I have been at the Uranium Institute, by how many of my colleagues in the nuclear fuel cycle are beginning to speak of the necessity of campaigns to right this situation, and indeed at the UI itself we have made a conscious decision to invest much more effort into public communication.

We speak of the importance of segmenting the targets of information work: of identifying the movers and shakers; of preparing long lists of the individuals who have a purchase on energy decisions, or who might influence the climate of opinion by what they write in journals of note. We consider politicians, including those on the way up, bureaucrats, journalists, lawyers and doctors, who often get asked by their patients about the dangers of radiation. Sometimes students creep into these target lists. They may be anti-nuclear demonstration fodder; they are also the decision makers of tomorrow, and some of them will be the nuclear engineers of tomorrow. This last group is one which we need to foster: in both Britain and the United States the falling off of the numbers of university students who opt to specialise in nuclear physics or nuclear engineering is a source of anxiety to the nuclear societies of both countries.

Technology for Global Economic and Environmental Survival and Prosperity
Edited by Kursunoglu *et al.*, Plenum Press, New York, 1997

131

Against a background where deregulation and increasing competition in the domestic electricity markets are forcing nuclear utilities to examine their overheads and to cut their expenditures to the bone in order to remain competitive, I wish to stress the importance of retaining an involvement in science education in general, and in the propagation of nuclear science in particular. This is not simply to maintain the supply of adequately qualified new entrants into the profession, but because it will have a generally beneficial effect in the long term on the public acceptability of nuclear power. We neglect education at our peril.

We can take heart from the fact that people are born without prejudices. A propensity to believe green propaganda is not an innate characteristic. Children are not born loving windmills, or energy-conservation. They learn to do so because anti-nuclear campaigners and environmentalists have made sure that such ideas are part of the common currency of the teaching profession in many countries. Casual observation suggests that small boys adore loud and heavily polluting racing cars or fighter aircraft, and the kind of technology-sparse world favoured by environmentalists is not one which the average man willingly embraces.

The low rating of nuclear science echoes that of science and engineering in general, but particle physics still has cachet: astrophysics and atom smashing at CERN still attract headline attention even in popular newspapers, but one does not read much there about new designs for nuclear reactors. Instead, the nuclear news is dominated by luddites who delight in every minor story about mechanical failure. We ought to ask ourselves why our technical breakthroughs no longer command attention. At the recent annual Symposium of the Uranium Institute an excellent lecture by Professor van Geel of the Transuranic Institute in Karlsruhe on the possibility of consuming plutonium in an inert matrix in a power station attracted no attention outside the nuclear industry. Given the popular but erroneous belief, inculcated by anti-nuclear campaigners, that plutonium is the most toxic substance known to man, and that we should look for ways of getting rid of it, this neglect is astonishing.

In contrast, doubtful theories about the connection between nuclear installations and leukaemia clusters are widely discussed. Could it be partly because of the much greater attention given to medical science in schools? People know about cells, blood and the micro-causes of disease. They also suffer from diseases and go to the doctor, so it is part of their direct experience too; but in developed countries at least, so is the use of electricity!

I do not purport to be an expert in educational matters, but I have consulted some who are and from them I have learnt some basic rules which need to be applied if we are to effect an improvement in our current situation.

The first rule is that whatever is proposed by way of a change in teaching materials has to be in line with the national or local curriculum. So where, as in England and Wales, there is below the secondary school level no niche in the compulsory part of the curriculum for the teaching of energy matters, still less of nuclear matters, it is difficult to get teachers to take them up. There are too many other claimants for the few optional slots which are left. This means in practice that in England and Wales very little discussion of energy takes place before a child reaches the 11–14 age group. But much younger children are exposed to atoms and molecules in their leisure reading, and an opportunity is being missed, as younger children can be shown to be much more open to innovation and the widening of their horizons than children of 11–14, among whom peer group pressure to conform tends to reduce their adaptability.

The second rule is that the aim should be to open minds rather than to convert them to a particular view.

The third is that teachers will accept carefully prepared material from any source, even one which they may regard as tainted (such as the nuclear industry) provided that the source is clearly declared, and the material is not propaganda. I understand that the Science and Technology Society in the United States has great success among American teachers by providing materials of this kind. BNFL likewise put considerable resources into the preparation of general and specifically nuclear materials for use in science teaching. The key is a willingness to find out what the needs of the children and the teachers are, and to provide it.

Teachers need to be convinced of the value of the subject. They will also be much better equipped than you or I to advise on the most effective methods of getting the message across. I had thought that the Uranium Institute might produce a bright poster for school use on the nuclear fuel cycle, to be brought up short by the advice of a Scottish head teacher that seven year olds are no longer impressed by such things these days, unlike twenty years ago, because much more exciting interactive or multimedia devices are now commonplace.

Fourth: the nuclear industry has much to learn from other industries who are in a similar position of being a public pariah. The chemical industry has almost as bad a press as the nuclear industry, but the divorce between this miserable reputation and the public's enthusiastic use of its products could not be more marked. When deregulation arrives nuclear utilities will need to learn to emulate the chemical industry's example.

In our approach to the education target we have two aims which might seem mutually contradictory, but which are in fact complementary. We need to recover the excitement of the early pioneers. Anyone who has read Rhodes book on *The Making of the Atomic Bomb* will know what I mean (some of the personalities concerned belong to this Foundation). The nuclear fuel cycle may now be a set of mature technologies, but it has yet to have its Frank Whittle. It is in the position of the civil aviation world before the arrival of the commercial jet engine. There are huge developments waiting to be made. Just as the commercial jet made mass travel commonplace, the nuclear electrical generating industry should be looking for comparable ways of fulfilling its early promises. After all, one third of the world's population (2 billion people) still has no access to commercial electricity. The global market is not saturated, whatever the followers of Amery Lovins might say.

Moreover in a world becoming agitated about atmospheric pollution and possible climate change, the ability of nuclear to deliver vast quantities of base-load electricity, even from its present technology, without the pollutants or the CO_2 which are the source of the agitation, is surely something which should be a cause for excitement if we set about it in the right way. The potential of the breeder reactor as the lynchpin of a clean, self-perpetuating energy cycle has foundered among a public who learnt at school that sodium has a violent reaction if exposed to air or water, but who did not learn that presently assured uranium resources would provide us all with electricity for many centuries to come if the Henry Ford of breeder reactors could be encouraged to appear.

Secondly we need to make nuclear technology seem as normal and routine as possible. In other words slightly boring. To achieve this we need also to persuade school children that nuclear science can be seen as ordinary, part of the natural world with which we are all familiar. Ionising radiation is like sunlight after all: a little of it does no harm and might even do some good (if the hormesis theory of radiation is soundly based); too much of it makes you ill; and a great deal too much will kill you.

To achieve this requires well thought out programmes applied systematically on a universal basis. Encouragingly I have discovered that there is a great deal going on in different parts of the world. I wonder how far the practitioners are aware of what their colleagues in other countries, or even in the same country are doing. I attach an annex giving some detail about a number of programmes which I have come across in researching this paper. They include a nation-wide British video based programme called Making Sense of Science, some work in teacher training colleges financed by the EU Commission, schools material circulated by BNFL and Nuclear Electric, Energy Centres in Japan, an excellent book for first year university students by the Professor of Physics at the University of Virginia, and work done in Canada by AECL and a small advocacy unit called Uranium Saskatchewan.

A point of great importance emerged from the EU work in South London. The radiologists involved found they needed to teach the teachers the science they hoped to impart to the children.

Teachers, particularly of pre-secondary schoolchildren, tend to be generalists, and therefore probably without much in the way of science background themselves. This is a danger, in that it predisposes them to be anti-nuclear; and from Britain to Japan they are frequently anti-establishment in political inclination, and thus temperamentally opposed to a big business/large-scale industry like nuclear. But it is also an advantage. In the absence of other resources, if the nuclear industry provides them with the materials on which to base their lessons these become the substance of what is taught.

Another point which I learnt from the Scottish head teacher (to whom I referred earlier) is that unless the subject is in the curriculum we will all be wasting our time showering the schools and teachers with material, however well crafted. This is perhaps more a public communications point than an educational one, but in segmenting our targets for improving public acceptance, the controllers of curricula, local and national, should figure high on the list.

CONCLUSION

To pull all these threads together, I began this paper thinking that we have found it all too easy to neglect the next generation. Nearly all our discussions at nuclear conferences are about short term problems: the present economic competitiveness or otherwise of the nuclear option, how to deal with public concern about waste disposal, when will the next station be ordered and so on.

The argument for taking a longer term view is a simple one. While it is probably true that as far as current decisions are concerned which affect the viability of the nuclear industry, in most countries they are taken against the background of the public acceptability or lack of it of nuclear energy. In most countries these days, even countries which have a tradition of accepting government decisions in a relatively conformist way, the nuclear power industry does not enjoy a high level of public acceptability. In the short term it is possible to tackle the situation by devising a determined campaign to influence the movers and shakers, those involved directly in the decision-making process, and those who are employed to influence the wider public in favour of the decision which has been taken. This may be a necessary condition for success, but it is not a sufficient condition. The political class may be persuaded, but if it is electoral suicide, or if it is perceived to be electoral suicide (a slightly different proposition) to carry out such a decision, then it won't happen. It is therefore essential in the longer term to create a situation in which the public acceptability of nuclear energy is greater than it is at present.

The same is true of its financial acceptability. The markets have to be convinced in the first place that an investment in a nuclear enterprise will deliver a good return on the capital employed. But if they perceive that the rules may be stacked against them because of public lack of acceptance they will not push the economic case. This has been the situation for a number of years at the World Bank, which has dodged the political hassle that a nuclear loan would have caused by finding economic excuses for not making one. (There is some evidence that the onset of market economics in the electricity supply industry and the increased worries about global warming have begun to make inroads into this rigidity.)

We need to consider how we can improve matters in twenty years' time. How do we ensure that there are enough qualified recruits to the workforce then? What about the world view of the opinion formers or decision makers then? The answer can only lie in education. It is easy to say that this is not the concern of the nuclear fuel cycle industries, and that it is the business of teachers and other educators. But we ought to take an interest in what the teachers teach, and to give support to the implementation of curricula which will benefit us in the medium to longer term. In other words, I believe that education is a long term investment. If the nuclear industry does not invest in education, and thus fails to demonstrate its commitment to the long term, how can it expect others to accept its statements about the longer term benefits of nuclear energy?

A great deal is already being done round the world, contrary to my initial impressions, as I have attempted to show, but its application is patchy and recent, and I am not sure that there is much cross fertilisation. Perhaps what is needed now is for the nuclear fuel cycle industries to establish an educational body with the same aims *pari passu* as the World Association of Nuclear Operators: the creation of an international code of good practice in nuclear education by the mutual sharing of experience and success.

One of its first aims should be to make the basic science more accessible. In such a campaign the teachers at all levels play a crucial role. It is they who teach the basics of science to primary school children, elaborate on the achievements of the earlier years in school when the children reach secondary/high school, and lead them through university or technical college – in some cases – to careers in the civil nuclear industry. (Not everyone is an autodidact like myself.) I would suggest that it is at the teacher training level that the nuclear industry could break out of the vicious circle which has entrapped it in an anti-nuclear public environment. It is there that the provision of appropriate teaching materials, scholarships and grants could begin to make an impact so that when the present generation of nuclear power plants come to the end of their planned, or even their extended planned life, the replacement investment will go into nuclear plant and not into some other means of generating electricity. The six-year olds of today are the university students of tomorrow, and the movers and shakers of the day after that, and while not all of them have such a glamorous future they will all be voters in 2008 when some of these decisions will be taken.

ANNEX

The British have had for the last few years a nation-wide schools programme called Making Sense of Science. It is well managed. It has involved school headmasters, heads of science departments and regular teachers. There is a sophisticated feed back programme attached so that the authors can learn the impact of the programme. The results are encouraging. While nuclear science is not entirely absent, it is not very prominent in this programme. We could push to have its role increased.

In a pioneering study Professor John Lakey and some colleagues attempted the creation of schools materials on radiation with the collaboration of a local education authority and a teacher training college in south London. The project was financed by the European Commission. As you might expect, the eminent experts in radiological protection had great difficulties at first in devising material which successfully put across what they wanted to express to primary school children, but with the aid of the trainee teachers and the children they eventually succeeded. They also learned a great deal about the receptivity of children at different ages: six year olds were much more open minded than ten year olds who for fear of peer group pressure are much more conformist. They also found they needed to teach the teachers the science they hoped to impart to the children.

Education is not confined to schools. Among others, the Japanese electrical utilities have shown us an excellent example of taking the education to the people. Many power stations now have excellent visitors centres which explain their technology. In Fukuoka the Kyushu Electric Power Company has built an excellent Energy Centre in the middle of town which succeeds admirably in showing that nuclear energy is part of the natural world. The Osaka Science Museum has much the same objectives, though in my view it is less successful. But everything which demystifies nuclear power is grist to the mill.

In the United States, the American Nuclear Society makes use of the expertise of many of its 14,000 members in educational activities with teachers and schools. In April 1995 the NEI ran a programme with its utility members to demonstrate to the local populations the environmental achievements of nuclear plants.

In Canada, AECL has been running a programme for some years in which it provides teachers with a few weeks of placement in the nuclear industry to enable them to gain a better understanding of the benefits and safety of the nuclear industry. AECL has found that this teacher placement programme has done much to overcome the lack of knowledge which has often led to utterly negative attitudes to nuclear, and has given the teachers concerned the confidence to provide their students with accurate information in a more positive light. This has had a beneficial effect all round, leading to an increase in the science and technology programmes available to students and teachers, an increase in the numbers of students choosing science or mathematics courses, an increase in the numbers of qualified persons entering the nuclear workplace, and a positive change in the attitude of local communities towards science and technology.

In Saskatchewan (Canada), which produces about 30% of the world's uranium, the provincial government includes in its school curriculum a unit on uranium and the nuclear industry. The province's uranium producers used their industry association, Uranium Saskatchewan, to exploit this educational opportunity.

Uranium Saskatchewan has a mobile information unit that travels around the province visiting schools. The unit, which consists of a 40ft trailer, contains static and interactive displays which explain what uranium mining is, the contribution that it makes to the province, and how it fits into the international nuclear industry. Before visiting a school, contact is made with the teacher(s) responsible and a teachers' guide is sent out. The aim is to meet all the curriculum requirements.

The uranium industry also funded a video on uranium in a series on Saskatchewan's resources. It provided the technical advice for the production and made a major contribution to the script. It has regularly organised visits to uranium mines by teachers, and provides speakers to schools on request.

As a result of Uranium Saskatchewan's campaigns public acceptance of uranium mining in the province rose by a third (from about 60 to 80%).

THE EAGLE ALLIANCE: A GRASSROOTS PROGRAM TO REVITALIZE NUCLEAR SCIENCE AND TECHNOLOGY IN AMERICA

Raymond W. Durante

President, Durante Associates, Inc. and Vice President, Eagle Alliance
1925 North Lynn Street
Suite 725
Arlington, VA   22209

Ladies and gentlemen, it is certainly a pleasure to be here with you and the Global Foundation meeting on the important topic: Technology for the Global Economic Survival and Prosperity.

In reviewing the program, I found it interesting that prosperity and survival have been linked with energy -- particularly with nuclear energy. While many in this room understand that linkage and agree it exists, most of the general population do not know (or care). Energy means two things: electricity that comes from the switch on the wall and gasoline that comes from the pump. In fact, these commodities are no longer the technical marvels we once thought but have taken on the status of entitlements in a class with air and water, and we go charging into the future convinced they will always be available in whatever quantities needed.

Not true. To ensure the availability of energy needed to "survive and prosper" as our conference theme states, one must plan carefully and, make the right decisions and follow through with the right programs. No one can accuse the U.S. of doing that -- we seem hell bent on burning the last cubic foot of natural gas and the last drop of oil just as fast as we can.

At an international meeting last June sponsored by the Center for National Policy in Washington, I chaired a session where Dr. Bill Gouse of the Mitre Corp. and former White House Science Advisor, gave an analysis of the world energy demands based on a number of studies conducted by various acknowledged experts. Except for the time period, all reached the same conclusions: based on projected growth of the third-world and underdeveloped countries of the world, we will run out of fossil fuels somewhere between 2100 and 2150 -- a not too distant 150 years away. Dr. Gouse reached this conclusion by tripling the known and estimated reserves of fossil fuels just to be sure there would be no argument. At the same time, the same experts agreed that use of electricity will continue to increase steadily just as it has since it was invented.

What will we use to produce this electricity if not fossil fuels? Interestingly, most predictors say it will be nuclear power even though it is dead in the water now and no utility includes the addition of nuclear in their future plans. Nuclear energy is the logical long-term choice and if we think in terms of 50 years or more, there is no other

Technology for Global Economic and Environmental Survival and Prosperity
Edited by Kursunoglu *et al.*, Plenum Press, New York, 1997

137

choice. The catch is that here in the U.S. we think in four year sound bites and sometimes not even that far ahead. In fact, our industry with the constant threat of mergers and takeovers is forced to look at quarterly reports and instant results. Despite the fact that we agree that we will need nuclear power in 50 or 100 years from now, we may not have a viable supply industry to meet these needs. But never mind, we can always buy the equipment from the Japanese, the French, or the Koreans and the lights will stay expensively lit.

But even if that is true and we are able to keep the lights on without nuclear power, there is another serious consequence we must consider. During the 50 years Americans resist using nuclear power, there will be a strong tendency to disregard "all things nuclear" and thus be deprived of the some of its additional benefits to mankind -- like nuclear medicine, food irradiation, industrial measurements, scanning, and fire and smoke detection. Hundreds of other applications derived from nuclear science and technology are in danger of being lost due to the public's fear of all things nuclear. Fear which is based on either lack of information or availability of the right kind of information.

That is a pretty good definition of education, which is the topic of this panel **Educating the Public**. In no case that I can think of is it more important to educate the public than in the case of nuclear technology. But we have not done a very good job and it will cost us dearly.

Enter the Eagle Alliance. In June 1995, Dr. Alan Waltar, then President of the American Nuclear Society, put forth the premise that we could not allow public opinion to limit the use of a technology that was so important to progress and social welfare simply because it was misunderstood and misrepresented by opponents. If this could happen to nuclear then it could happen to any similar advanced technology that had some risk associated with it. He prepared and distributed The Declaration of Energy Interdependence, a document that encourages nuclear professionals to join together in an alliance to help preserve the progress of peaceful uses of nuclear technology and ensure that its benefits are available to all humanity. This document was signed by more than three dozen leaders in the medical, industrial, academic, and government fields who share the belief that now was the time to move forward.

This impetus led to the formation of the Eagle Alliance, a not-for-profit educational organization whose mission is to provide accurate information to the public and decision-makers regarding the application and impacts of nuclear science and technology. The American Nuclear Society, the American Medical Association, the Institute of Nuclear Physicians, and the College of Radiological Pharmacists are founding members of the Eagle Alliance. We began with a small group working on their own time and mostly through teleconferences. Eventually we established headquarters in Washington, DC in an office shared with the Association of Engineering Societies. So we far have conducted three public meetings in Washington DC. All this has been done on a volunteer pro bono basis with many individuals devoting their time and efforts to developing and implementing start-up procedures. The American Nuclear Society and the Nuclear Energy Institute have been very helpful in providing some money and services in-kind for materials and operating expenses, and several companies have made cash donations to help with the costs of mailing and conducting meetings. We do not, however, want to be just another trade association and although we have three classes of membership (corporate, institutional, and individual), we hope to have the overwhelming participation be by private citizens. We are now in the process of soliciting individual membership and expect to have more than 1,000 members by early spring of 1997.

At this time I am sure a number of questions have entered your mind -- like hasn't this been done before? How are we different, industry has spent millions trying to do the same thing and failed why will we succeed? Why would ANS and NEI help us -- aren't we in direct competition to them? How will we "educate" the public? Are we a lobby -- trade association? Where will we get money to operate? Do we have a staff? Who? What? Why?

I cannot cover all of these questions in the time I have left, but would like to address some of the more important ones. While we share many of the same objectives of other nuclear organizations now operating, our approach to the problem will be quite different. For example, the NEI has emerged as a strong voice in behalf of nuclear technology, particularly with Congress and selected government agencies. They focus primarily on <u>issues</u> and since their support is solely from industry, they must concentrate on activities beneficial to their members welfare. An example is the high level waste disposal problem which has dominated their activities for the past three years. The ANS, on the other hand, strives to be the ultimate unbiased source for scientific data and tends to avoid controversial issues since it is difficult to achieve a consensus among its 15,000 members. The Eagle Alliance fits neatly in between and compliments each one while doing its own thing. The Alliance aims to be a grassroots organization, including both industry and the scientific community, but not owned by either one. Much of the success enjoyed by our opponents (i.e., Greenpeace, and NRDC) has been due to the use of grassroots techniques -- the use of "ordinary citizens." Somehow we believe things more readily if said by our neighbor, our barber, or family member and look askance at professionals who may have parochial interests. On the other hand, there are a lot of people in our industry who fit the category of "ordinary people" whose voices are never heard because there is no vehicle to put forth their viewpoints.

Nuclear technology represents a $330 billion industry with 4.1 million jobs. There is a vast army of blue and white-collar workers involved in nuclear technology who know firsthand this technology is safe, reliable, and beneficial. Everyone in the medical profession, from the floor nurse to the chief administrators, knows the contribution made by nuclear medicine to the welfare of their patients. They also know the dilemma faced by the medical profession because of public resistance to low level waste sites. Nuclear plant construction workers are aware of the safety procedures and care required in building facilities and can refute the false claims made by anti-nukes regarding the safety of these plants. Thousands of industry and national lab employees, either on the job or retired, can cite the advantages and benefits of spin-off technology from nuclear science and technology projects, and they know first hand of the extensive resources and talents available to solve almost any problem that arises. These are but a few of the resources which, if properly directed, can be an effective truth squad and carry the right messages to the general public. The Eagle Alliance will seek the help of all of these people by mobilizing them and providing information and data they can understand and relate to others.

With regard to the question of whether or not Eagle Alliance will be a lobby organization, the answer is definitely no. We are a 501 (3)(c) corporation whose main mission is education, promoting awareness of the benefits of nuclear technology. We hope the messages crafted by the Eagle Alliance will find their way into the hands of the decision makers at the local, state, and federal levels. In today's fish bowl environment, the definition of lobbying has been fine tuned to the point where anyone even having a conversation with an elected official is suspect. That is unfortunate since it is impossible for any legislator to make decisions in a vacuum, especially on complex technical matters. Today's legislators make decisions based on knowledge gained from staff, constituents, and the media. They will be our targets and, if we approach them in a reasonable and constructive manner, we should be well received. I think most of you will agree, nuclear technology gets no help from the electronic or written media, and the science fiction writers and Hollywood elite have a field day at our expense. Any counterattack by industry is branded parochial and self-serving and dismissed summarily. Responses from an organization such as the Eagle Alliance stand a better chance of getting a fair hearing.

This situation would be corrected or at least mitigated with a pro-active Administration or politically independent technical agencies who would be willing to present the facts -- including risks verses benefits to the public. This Administration not only opposes nuclear but coal as well and as long as we have cheap natural gas, they will not have to defend their position. Therefore, messages directly to the President and his Cabinet, while not too effective must continue, if for no other reason than to let them know we do not agree with their position. Opponents of nuclear technology are determined to eliminate "all things nuclear" and have approached their task with an almost religious zeal. To respond with scientific facts and risk/benefit ratios is no use and in most cases is distorted when played back. The only way to respond is to match their numbers and stay with the truth.

An excellent case in point is the food irradiation story. This technology has been available for more than 30 years, is proven safe beyond scientific doubt, accepted by every major regulatory body and government health agency and used in 37 countries throughout the world. It has the potential to protect the health of our citizens and increase the world food supply. Yet it is not in widespread use in the US and food suppliers fearful of criticism and even boycott will not use it. As a result, there are 81 million cases of food sickness each year and 25 percent of the world's food supply is lost to spoilage and infestation. Both problems could be significantly mitigated through widespread use of food irradiation. In this case the public must not only be informed of benefits being denied to them but must DEMAND they be made available.

First and foremost we can help disseminate the truth about the benefits derived from nuclear technology (i.e., MEDICINE, FOOD, ENERGY, INDUSTRIAL, AND OTHER). Each of these areas have valid and not so valid concerns that have been emphasized by opponents to "all things nuclear" that must be addressed. For example, where do we get the radionuclides needed for treatment and diagnosis? Do we have the necessary safe disposal facilities? How much evidence has been established to prove these processes are safe? Why doesn't our government support these beneficial technologies? Why have utilities stopped building nuclear power plants? Why have the major US companies abandoned nuclear?

We have to have to anticipate the questions and have answers ready that include irrefutable facts. On the other hand, we must be careful to not overpower the public and deluge them with scientific data they cannot comprehend. All too often the phrase "educate the public" is taken literally and many of our colleagues try to teach courses in nuclear technology which can result in more fear and doubt than reassurance. Take the case of the microwave oven. When this appliance first emerged on the scene there was much public concern and its use was relegated to commercial behind-the-counter use. Now, almost every home has one and convenience stores and cafeterias allow their unsupervised use by young and old alike. The PERCEIVED AND REAL danger is still there. Why aren't we concerned the "uneducated" public will harm themselves? They still don't know how the thing works and no one ever tried to tell them. The product was sold and accepted on the basis it offered enormous benefits with very low, manageable risks.

Reliable polls showed that a broad cross-section of the U.S. public supports Nuclear Technology and it's applications, but they get hung up on the tough questions. Where do we put the waste? How safe is safe? Why is the NRC levying so many fines? If the Eagle Alliance can become a reality, we hope to provide answers to these questions in a way they can be understood and believed by the ordinary voting citizen and politicians.

In addition to providing answers to questions and being continuously on the defensive, we have to go on the offense as well and convince people about the benefits of nuclear science and technology rather than teach them how a nuclear reactor works. For example, some of the things they should know are that 13 million nuclear medical procedures are done every day every year and nuclear diagnosis or therapy benefits 1 out of every 3 hospital patients. Nuclear sterilization assures the safety of surgical tools and bandages. Nuclear tests are used by the FDA for approval of over 80 percent of all new drugs. The mileage rating on automobile tires is doubled after exposure to ionizing radiation. The food taken by astronauts into space and the meals hospitals serve to organ transplant patients are irradiated to ensure there is no transmittal of disease or harmful pathogens. Food irradiation may be the only way to destroy E. Coli in red meat and increase shelf life of fresh produce and seafood helping to fight world hunger. Nuclear plants supply 20 percent of all electricity produced in the United States and represents the only large scale, sustainable fuel source with minimal pollution. The 40 year safety record of the nuclear utility industry is unsurpassed. They should also know that hundreds of nuclear applications in sensors and isotope tracers, thickness gauges, smoke detectors, and airport check-in devices.

Reliable polls consistently show that the public, when advised of the true facts concerning nuclear science, are much more favorably inclined towards its support and development. So it is our hope that the Eagle Alliance can be the vehicle to provide that information. The question remains what can be done and how can we do it? We can encourage local schools to include accurate information on nuclear technology in their curriculum. We can assist the media in providing balanced and accurate reporting and demand that special interests and activists organizations be held accountable for the truth and accuracy of their claims. Citizens in a democracy require adequate knowledge in order to govern effectively and, therefore, we should do all we can to see that those in positions of governing get the proper and full information. We have to work towards maintaining US leadership in domestic and international nuclear affairs and do what is necessary to encourage our institutes of learning to include nuclear curricula as part of their basic education.

More specifically and in the immediate future, we intend to produce brochures describing medical nuclear applications and put these brochures in every doctor's office. We expect to hold national and local meetings and allow our members to freely exchange information and set up chapters and local and student sections. We plan to produce literature as hard copy and on a Web page and e-mail and challenge the media to report more accurately.

I brought some descriptive literature on the Eagle Alliance with applications for membership. We need to raise funds. We need volunteers. We need your help. I hope you will all give this serious thought. Thank you very much for your attention.

NUCLEAR EDUCATION ON POLITICAL AND MILITARY STRATEGIES[*]

Behram N. Kursunoglu

Global Foundation, Inc., Coral Gables, Florida
(kursungf@netrunner.net)

A course motivated from the descriptive title of the commencement speech in the Florida Institute of Technology was proposed by me and was approved by various academic committees of the University of Miami. It had received enthusiastic support from all, and three credit course entitled "Nuclear War-Nuclear Peace" was scheduled for spring term of 1983. One of the novel aspects of the course was the fact that its proposed contents included a broad spectrum of points of view: scientific, technological, strategic, military, political, and global issues as they affected the subject matter. This approach was achieved by having visiting lecturers with well established track records in their own specialties. The course included diverse points of view of some world class scholars.

The proposed course description as shown here was not followed exactly but effort was made to include the discussions of all the subjects listed. An interdisciplinary course on Nuclear War-Nuclear Peace was sponsored by the Center for Theoretical Studies in cooperation with the Departments of Politics and Public Affairs, Sociology, History, Civil Engineering, Physics, and Programs in Honors and Privileged Studies, and was offered to undergraduates and graduates as a three-credit course. The above departments had listed it in their curricula as a 200-level course.

I. INTRODUCTION

1. History of the Manhattan Project 1939-August 6, to 1945

2. The Impact of the Nuclear Weapons Monopoly on the Post World War Scene (1945-1950)

3. Transition into the Breaking of the Nuclear Monopoly and Beginning of a World With Two Superpowers

Five films:

The Tale of Two Cities and *Hiroshima and Nagasaki*. The first film, made by the U.S. Army, shows structural damage but no human casualties. The second film, made by a peace group, shows people. The contrast is excellent.

Countdown for America, by the American Council for National Security.

The Last Epidemic, by the Physicians for Social Responsibility.

The Day After Trinity, by Pyramid Films.

---

[*] This paper is a verbatim from the typescript of the author's book with the provisional title "The Ascent of Gravity: Einstein's Triumph, Paradigms, and Contemporary Physicists" which is to be published in 1997 or early 1998.

Recommended Readings:

(Publication information to be supplied where needed)

H. D. Smyth, 1945, Atomic Energy, Washington D.C.: U.S. Government Printing Office; London: HMSO.

Robert J. Art and Kenneth N. Waltz, eds., The Use of Force, (article by Morton "The Decision to Use the Atomic Bomb").

Gregg Herken, The Winning Weapon, 1982.

B. Brodie, 1947, in the Absolute Weapon, New York, Harcourt Brace, pp. 21-110.

II. Impact Of Basic Research On Military Industrial Complex

1. The physicists Who Began the Nuclear Era From 1939-1950: Albert Einstein, Leo Szilard, Robert Oppenheimer, Hans Bethe, Enrico Fermi, Edward Teller, Eugene Wigner, Richard Feyman...

    Other physicists involved: Bruno Pontocorvo, Klaus Fuchs, and Andrei Sakharov...

2. Elementary Concepts of Physics Involved in the Construction of a Nuclear Weapon

3. The Beginning of the Nuclear Era
    a) Nuclear energy for the production of electricity
    b) US-USSR nuclear balance and peace

4. The Problem of Nuclear Waste (Civilian, Military)

Recommended Readings:

(Publication information to be supplied where needed)

S. Zuckerman, Scientists and War, 1966, London: Hamish Hamilton

J. S. Foster, "Nuclear Weapons," Encyclopedia Americana, 1972, Volume 20, pp. 518-528.

Georges A. Vendryes, "Superphenix: A Full-Scale Breeder Reactor," Scientific American, March 1977.

A. Perlmutter, et al, editors, Nuclear Energy and Alternatives, Proceedings of the International Scientific Forum on an Acceptable Nuclear Energy Future of the World, Ballinger Publishing Company, Cambridge, Massachusetts, 1977.

B. N. Kursunoglu, and A. Perlmutter, editors, Directions in Energy Policy, A Comprehensive Approach to Energy Resource Decision Making, Ballinger Publishing Company, Cambridge, Massachusetts, 1978.

Ralph Lapf, A Citizen's Guide to Nuclear Power,...

B. N. Kursunoglu, and A. Perlmutter, editors, Impact of Basic Research on

Technology. Plenum Publishing Corporation, 1973.

III. Nuclear Weapons As A Deterrence For The Free World 1956-1970

1. 1962 Cuban Missiles Crisis and Confrontation with the USSR

2. The Benefits of Nuclear Weapons for Peace!

3. The Beginning of the Vertical Proliferation of Nuclear Armament
   a) role of the hydrogen bomb
   b) ballistic missiles nuclear fire carrier

Recommended Readings:

The Role of Technology in Meeting the Defense Challenges of the 1980's

H. F. York, Race to Oblivion, 1970, New York: Simon and Schuster

IV. The Four Theaters Of Nuclear War

1. Land-Based Ballistic Missiles

2. Nuclear Submarines Carrying Nuclear Weapon-Loaded Missiles (antisubmarine warfare)

3. Missiles in the Air (cruise missiles)

4. The "Nuclearized Space" (antisatellite weapons, and laser and particle beam weapons)

5. (C3I) (command, control, communication, and intelligence)

Recommended Readings:

"The Effects of Nuclear War," 1979, Washington, D.C., Office of Technology Assessment, U.S. Government printing office

Richard Garwin, "Antisubmarine Warfare and National Security," Scientific American, July 1973

Kosta Tsipis, "Cruise Missiles," Scientific American, February 1977

Richard L. Garwin, 1981, "Weapons in Space: Are we on the Verge of a New Arms Race?" (paper presented to AAAS Annual Meeting, Toronto, 3-8 January)

V. Fundamental Problems In The Guidance Of Nuclear Weapon Delivery Systems

1. The Problem of Electromagnetic Pulse

2. Optical Fibers in Space Communications

3. Accuracy in Ballistic Missile Guidance and Laser and Particle Beam Weapons.

Recommended Readings:

Richard L. Garwin, 1980, Bureaucratic and Other Problems in Planning and Managing Military R&D, in Long & Reppy, The Genesis of New Weapons, London: Pergamon

John Parmentola and Kosta Tsipis, "Particle Beam Weapons" in Scientific American, April 1979, 240 (4), 38-49

VI. Nuclear Deterrence, Nuclear Freeze, No First Use, Mutually Assured Destruction (Or Mutually Assured Survival)

1. The Concept of Nuclear Fire Power Parity

2. First Strike

3. Second Strike

4. Launch Under Attack

5. Can One Win a Nuclear War?

Recommended Readings:

Sir Solly Zuckerman, 1982, Nuclear Illusion and Reality, The Viking Press, New York

"European Theater of Nuclear Forces," The Bulletin of Atomic Scientists, October 1980

Sidney Drell and Eric von Hippel, "Limited Nuclear War," Scientific American

R.W. Clark, 1980, The Greatest Power on Earth, London: Sidgwick & Jackson, p. 126

VII. US-USSR Nuclear Arms Reduction Prospects

1. Problems of Verification

2. Impact of Conventional Weapons on Nuclear Armaments

3. Impact of Horizontal Nuclear Arms Proliferation on US-USSR Negotiations and Vice Versa, Who Else Has and Will Have the Bomb?

4. A List of Scenarios to Start A Nuclear War

Recommended Readings:

Sidney Drell, "Arms Control: Is There Still Hope?" Daedalus, Fall 1980

Barry Blechman, "Do Negotiated Arms Limitations Have a Future?" Foreign Affairs, Fall 1980.

VIII.   Effects Of Nuclear Weapons

    1.   Stratified Circles of a One-megaton Nuclear Explosion on a City

    2.   Casualties from Shock Waves, Radioactive Fallout, etc.

    3.   The State of Survivors in a Post Nuclear Explosion

Recommended Readings:

"Effects of the Possible Use of Nuclear Weapons," 1968, Report of the Secretary General to the United Nations, New York. See also the 1980 report of the same

"The Effects of Nuclear War," 1979, United States Arms Control and Disarmament Agency.

Most visiting scholars for the course have lectured to the public on Wednesday nights on the same subjects. Some of these evening lectures as happened during one of the Edward Teller Wednesday night lectures, did experience disturbances by a few crackpots among the audience. Coral Gables City Police had intervened by making a few arrests right on the spot. Teller had participated as a distinguished lecturer during the eleven years of 1983-1993, and in some of them he lectured to the public also. After the first incident whenever Teller lectured we made arrangements with the police to guard the lecture room. As an example to illustrate our evening lecture series I am including here the 1983 list and topics of lecturers.

UNIVERSITY OF MIAMI
CENTER FOR THEORETICAL STUDIES

Lecturers

Nuclear War-Nuclear Peace Lecture Series for 1983
University of Miami Main Campus
Wednesday evenings, 8:00 p.m., Room LC110

January 12, 1983   Prof. Behram N. Kursunoglu; Director of the Center for Theoretical Studies, University of Miami
"Critical Issues and Choices in the Nuclear Era"

January 19   Prof. Victor F. Weisskopf; Massachusetts Institute of Technology.
"How to Avoid Nuclear War"

January 26   FILMS - *Countdown for America*, by the American Council for National Security, and *The Last Epidemic*, by the Physicians for Social Responsibility

February 9   Lord Solly Zuckerman; Distinguished Zoologist, A Principal Science Advisor to the British Government
"European Perspectives on Security"

February 16   Prof. Eugene P. Wigner (Nobel Laureate) (member of the Manhattan Project); Princeton University
"Desirable Objectives of Government: Power or People's Happiness?"

February 23   Prof. Edward Teller; (member of the Manhattan Project), Hoover Institution, Stanford University
"Two Crises: The Persian Gulf and NATO"

March 2   Mr. Paul C. Warnke; former chief U.S. SALT Negotiator
"Strategic Arms Negotiations: The Present Prospects"

March   Prof. Warner R. Schilling; Director, Institute for War and Peace Studies, Columbia University
"American Approaches to the Nuclear Dilemma"

March 23   Mr. Joseph Lehman; Public Affairs Advisor of the U.S. Arms Control and Disarmament Agency
"Arms Control in the 80's--A New Approach"

March 30   Dr. Donald M. Kerr; Director of Los Alamos National Laboratory
"Nuclear Defense of Europe"

April 6	Dr. Dixy Lee Ray; former Governor of the State of Washington, and former Chairman of the U.S. Atomic Energy Commission "Nuclear Knowledge and Nuclear Power"
April 13	Mr. Alexander Zotov; Counselor of the Soviet Embassy "Arms in Europe"
April 20	Dr. C. Pierre Zaleski; Deputy-Director of the Center for Geopolitics, of Energy and Raw Materials—University of Paris - Dauphine "The French Perspectives on Nuclear Power and Nuclear Weapons"
April 27	Prof. Behram N. Kursunoglu; Director of the Center for Theoretical Studies, University of Miami "Dialogues with Albert Einstein and Other Physicists, and the Evolution of the Nuclear Era"

In its first few years offering of the course the number of students enrolled in the course varied between 200 and 280. After 1986 the numbers began decreasing and by 1993 there were only 47 honors students taking the course. In 1985 when Eugene P. Wigner was to lecture for the third time on the course he had a slight cold and laryngitis, asked me if I could repeat, over the loudspeaker, his lecture while he is giving it. I was very pleased to help him out with his lecture "Recollections of the Connections With the Uranium Project". However, Wigner was not quite aware of the fact that ninety percent of the 280 students had never heard of the science of physics and, therefore, I found myself doing two things at the same time: (i) try to remain as close as was possible to Wigner's overall talk, (ii) do not use any language of physics! Thus, I had to perform two acts to contradict each other. When Wigner's lecture was over I signalled the class, while he was not looking, to give Wigner a standing ovation. This they did and it was like a pageantry. Wigner was very pleased and he expressed it by saying that: "This is very nice, but you did not say what I said!" I replied by saying that: "Eugene you are right I did not say what you said, but that is why you got a standing ovation!" Wigner, "You have a share too". While I am relating anecdotes about Wigner, I could not resist the temptation to relate one more. One of our distinguished lecturer, as listed earlier, was Victor Weisskopf who was a close Wigner collaborator and they wrote some classic papers together. They were, once, attending some one's lecture. When they came out Wigner turned to Weisskopf and stated that the lecture was so simple that he could explain it even to a stupid guy. To which Weisskopf responded as "Therefore, explain it to me". Wigner smiled and said "Let us not say I can explain it to a stupid guy, but let us say that I can explain it to an average guy!". After his lecture to the class I dared to ask Weisskopf about this Wigner story to which he responded by claiming that "the word was not *average* but just *stupid*!"

Lord Solly Zuckerman, who passed away in 1993, was one of the trustees of the Global Foundation. Alexander Zotov, one time lecturer in the course, whose official title was Counselor of Soviet Embassy, reminds me another interesting occasion with Wigner. A few times I made arrangements to meet Wigner in the Washington, D.C. Cosmos Club. This time in 1984 we were joined by President Reagan's science advisor George Keyworth. When we left the club I was greeted at the door by Alexander Zotov who just came out of his car to inform me that academician Yevgeny Velikov was in the Club and Zotov would like me to meet Velikov, just at this time Velikov appears and when he sees Wigner he greets him by addressing him "Eugene it is so good to see you". But Wigner looked at me and asked, pointing to Velikov, "Do I know this gentleman?" After receiving the envelope containing an invitation to me and Sevda the Soviet Embassy reception in honor of academician Velikov for the next day we left Cosmos Club.

Other lecturers in the course included Lt. General James Abrahamson, Director of the Strategic Defense Initiative Organization, and one of his aids Col. Peter Worden; Gerald Yonas, Chief Scientist to SDIO who described our course as one of the best in the U.S.

Albert Carnsale (Harvard University), Ambassador Seymour Weiss (formerly U.S. Department of State). Hans A. Bethe (Cornell University, Nobel Laureate in Physics); Michael Burger, O'Dean Judd, Linda Trockie all three from Los Alamos National Laboratory, Joseph D. Lehman of USACDA and later the Martin-Marrietta Company. Robert Hofstadter of Stanford University and Nobel Laureate in physics was also a trustee of the Global Foundation, John S. Foster (TWR), Robert T. Grey (Deputy director of U.S. Arms Control and Disarmament Agency). Arthur Singer (Vice President of Alfred P. Sloan Foundation) is one of the outstanding experts in the 1962 Cuban missile crisis lectured in the course in the years 1989-1992, who was also a member of the Global Foundation Advisory Board; Henry King Stanford (President Emeritus of the Universities of Miami and Georgia) lectured the class in the years 1990-1993, is also a trustee of the Global Foundation; George Wald (Harvard University, and Nobel laureate, first mentioned in this book in the prologue, has had a different style in his presentation where he asked the students, before his lecture, to join him in breathing exercises which the students enjoyed greatly; Stanislav Ulam (Chairman of the Department of Mathematics, University of Colorado), one of the outstanding mathematicians of the 20th Century was involved in the design of the hydrogen bomb. He lectured the class in 1984 and spoke on his equations for the hydrogen bomb. He elaborated on how a set of complicated mathematical equations had to be handled with the old fashion computers of the 1940's. This was Ulam's second visit to CTS after his first visit in 1976.

It was then that my colleagues and I went to a fancy party in Miami where a young lady was quite fascinated with Ulam and she wanted him to comment on her see-through blouse upon which he said to her "I like what I see!" Stanislav Ulam and I, after his class, walked to the Faculty Club which was about one mile away from the classroom. On the way to the Club we talked about cosmology and he expressed his pleasure about the nice, cool winter day, as well as the beauty of the subject matter, we were discussing. I do not know the details and the politics of Ulam's work on the hydrogen bomb, but a week after Ulam's visit Edward Teller called to protest to me for the good reception we gave to Ulam especially on the subject matter of the hydrogen bomb. Luckily I was not in town so he decided to converse for about 30 minutes with my secretary Linda Scott on the history of the bomb as he knew it. He did not hide his disappointment with me even though I was not guilty of any wrongdoing. However, Edward Teller being a kind friend did not keep any ill feelings and he was soon back in CTS participating in the usual activities. Needless to say Linda was not a physicist, and Teller did not take that into consideration even though he knew her from his visits to CTS in the past 30 years. It was a great disappointment to learn, only two months after his visit to CTS, of the bad news that Ulam suffered a heart attack and died at the age of 75.

There was also appropriate occasions to invite a scholar from the Archdiocese of New York, Brother Austin David, ESC., Ph.D. He prepared two lectures to deal with the *Challenge of Peace: God's Promise and Our Response* which was the title of a letter prepared by the Catholic Bishops' of the U.S. to deal with the issue of war and peace. He titled his two lectures as *The Churches and the Nuclear Debate: Ethical and Moral Consideration*, and *The Challenge of Peace*. For his Wednesday evening public lecture he chose *The Fire of the Gods*. I had also invited His Excellency, Archbishop John J. O'Connor of New York. He, however, declined the invitation.

From the U.S. Navy Vice Admiral John Marshall Lee lectured the class on the two topics *What Use Are Nuclear Weapons?*, and *Changing Our Thinking in the Nuclear Age*. For his Wednesday evening public lecture he chose the topic *Thinking about Nuclear Weapons*. There were others who participated as visiting professors of a week duration. Franklin A. Long of Cornell University talked to students on the two topics *Unilateral Initiatives as a Useful Path to Arms Control and Disarmament*, and *A skeptical View of President Reagan's star Wars Proposal*. For his Wednesday evening public lecture he chose the topic *New Directions In Negotiations for Arms Control and Disarmament*.

Alvin M. Weinberg of the Institute for Energy Analyses, Oak Ridge Associated Universities was also amongst the visiting professors of one week duration. There were great diversity of viewpoints on this subject which can further be illustrated by listing Weinberg's two topics for the class *The Defense-protected-build-down*, and *New Approaches to Non-proliferation Policy*. The public lecture was on *The Successes, Failures, and Prospects of Nuclear Energy*.

George Rathjens and Jack Ruina from MIT gave lively examples where one thinks of the famous French saying "Vive la Difference!" For the Wednesday evening public lecture Ruina chose the topic *Star Wars: The President, The Prognosis, The Pitfalls*. Whilst for students the subject *The Nuclear Balance: Why People Judge It Differently* was presented in two lectures.

The eleven years (1983-1993) of offering the course was certainly worth while but an expensive enterprise. I had to raise the funds from various sources. A $20,000 grant from Alfred P. Sloan Foundation in 1985 was most encouraging for the future of the course. This grant also helped in the early years of the course to produce video tapes of the lectures both in the class and during the presentation of the public lectures on Wednesday evenings. A contribution of $67,000 to CTS from the Global Foundation's nuclear programs during the years 1983-1988 made it possible to continue the visiting professors participation in the course lectures. Overall cost of the 11 years lifetime of the course was approximately of the order of $150,000 which included modest honoraria of $200 per lecture, round-trip airline ticket, hotel room and $75 per diem.

Examples of the Course Schedule and a Final Test

Ninth Annual Offering of
NUCLEAR WAR - NUCLEAR PEACE

sponsored by

UNIVERSITY OF MIAMI'S
CENTER FOR THEORETICAL STUDIES

Spring Semester 1992
Three-Credit Course

Honors 331 & Physics 122

Tuesday & Thursday
3:05-4:20 p.m.
Learning Center - Room LC194

An Interdisciplinary Departmental Course

Lectured and Directed by Professor Behram N. Kursunoglu,
with assistance from Professors Arnold Perlmutter and Robert Kelly

The basis for the course Nuclear War-Nuclear Peace is the belief that an enlightened youth is the best security in a nuclear-armed world. The course (which is offered to undergraduate students at the University of Miami) will cover the development of the atomic and hydrogen bombs by the United States and the Soviet Union in the period from 1939 to 1955 and the influence that the existence of these weapons has had on military strategy and international relations. Special emphasis will be given to Gorbachev's role in the recent East European events as they pertain to the decline and collapse of Communism and to the role of nuclear weapons in maintaining world peace. The discussions will include the Persian Gulf Crisis and the possible use of chemical, biological, and nuclear weapons because of the proliferation of these weapons in the third world countries. The problems of energy for industrialized and developing countries, along with environmental issues and global warming will also be included.

The course utilizes a series of distinguished lecturers who have been directly involved in the events of the nuclear age. It includes physicists who were involved in the Manhattan Project and political scientists who have been involved in the creation of some of the recent treaties to control nuclear arms.

Tea Seminars will be held as announced so that small groups of students will be able to meet and discuss issues with the lecturers. These seminars will be held at the Center for Theoretical Studies, which is located in Building 21-V (second floor), 5606 Merrick Drive.

This three-credit, permanent course, is sponsored by the Center for Theoretical Studies and the Global Foundation, in cooperation with the Department of Physics and Programs in Honors and Privileged Studies. Students may enroll in PHY122 or HON331 to receive credit (3 hours) in the course. Special efforts for honors students include writing two short reports on any two presentations by course lecturers. Honors students are expected to attend at least three Tea Seminars during the semester.

CENTER FOR THEORETICAL STUDIES COURSE PROFESSORS INCLUDE

Behram N. Kursunoglu; University of Miami
Theoretical Physicist; Einstein scholar;
Professor of Physics; Director of the Center for Theoretical Studies

Michael E. Berger: Los Alamos National Laboratory
Program Director for Environmental Research
Applications Program OFFICE

O'dean P. Judd; Los Alamos National Laboratory
Formerly Chief Scientist Strategic Defense Initiative Organization;
currently Chief Scientist, LANL Director's Office

Joseph D. Lehman; Martin Marietta Astronautics, Denver
Director of Strategic Systems, Martin Marietta Astronautics;
Former Director of Public Affairs, U.S. Arms Control and Disarmament Agency

Arthur L. Singer, Jr; Alfred P. Sloan Foundation, New York
Vice President

Henry King Stanford; Americus Georgia
President Emeritus, Universities of Miami and Georgia

Richard J. K. Stratford; Department of State
Deputy Assistant Secretary of State for Nuclear Energy
and Energy Technology Affairs

Edward Teller; Hoover Institution, Stanford University
Theoretical Physicist; Member of Manhattan Project; "Father"
of the Hydrogen Bomb

On my honor, I have neither received or given aid on this examination

SIGNATURE:_____   I.D.:_____

University of Miami
Center for Theoretical Studies
NUCLEAR WAR-NUCLEAR PEACE
HON 331 - FINAL EXAM

A three-credit course for students from Program in Honors and Privileged Studies. Organized and sponsored by the Center for Theoretical Studies.

Led by Professor Behram N. Kursunoglu
With assistance from Professors Arnold Perlmutter and Robert Kelly

NUCLEAR WAR - NUCLEAR PEACE   May 5, 1992

[1]   Attempted proliferation of weapons of mass destruction in third world countries include
       [A]   Outer Mongolia, Switzerland, Nigeria, Guatemala
       [B]   Brazil, North Korea, Iraq, Trinidad, Bahamas
       [C]   Iraq, Syria, Pakistan, North Korea, Libya, Iran
       [D]   Israel, Ireland, Nigeria, Kenya
       [E]   none of the above

[2]   Missile accuracy, the degree of undetectibility of submarines (silence) in the ocean, and radar avoidance of detection of long range bombers are
       [A]   strategic advantages
       [B]   necessary technological edges for superpower status
       [C]   important deterrences
       [D]   all of the above

[3]   Which of the following requirements are needed for a country to acquire and maintain superpower status?
       [A]   largest military outlay exceeding its economic power
       [B]   largest economic power exceeding that of its military outlay
       [C]   very large economic power affording the largest military deployment
       [D]   all of the above

[4]   MIRVing of US ballistic missiles was the result of
       [A]   increasing nuclear bomb yields and missile accuracy
       [B]   limiting the number of missiles as agreed upon by USA and (former) USSR
       [C]   implementing the ABM treaty signed by USA and USSR
       [D]   implementing the LTB treaty signed by USA and USSR
       [E]   none of the above

[5]   The mechanism of energy production in the Sun and other stars is
        [A]   based on high temperature superconductivity
        [B]   based on the fission of heavy elements as in the uranium and plutonium
                bombs
        [C]   based on combustion as in the burning of fossil fuels
        [D]   based on the fusion of light elements as in the hydrogen bomb
        [E]   two of the above

[6]   The gradual reduction of the size of nuclear warheads by USA and (former) USSR
        was due chiefly to
        [A]   a growing shortage of plutonium
        [B]   economic pressures
        [C]   increased accuracy of ballistic missiles
        [D]   increased yields of nuclear bombs

[7]   Which of the following energy sources are renewable?
        [A]   fusion [B] coal   [C] fission (via Breeder) [D] natural gas
        [E]   solar (including also wind, hydro, biomass)   [F] oil

[8]   The time for an ICBM to reach Russia when fired from the central USA is about
        [A] 5 seconds [B} 5 minutes [C] 30 minutes [D] 3 hours
        [E] much longer than all of the above

[9]   The principal motivation for President Reagan to propose SDI in 1983 was to
        [A] replace MAD by MAS   [B] replace ABM by MAD
        [C] to extend a nuclear umbrella over all the Western allies
        [D] provide more work for a starving defense industry
        [E] bankrupt the Soviets

[10]   The acknowledged principal architect of the MAD strategy was
        [A]   Gerald Ford   [B] Andrei Sakharov   [C] Jimmy Carter
        [D]   Robert McNamara
        [E]   Bernard Baruch

[11]   A geosynchronous earth satellite
        [A]   orbits the earth with the same period as one earth day
        [B]   will be used mainly to emit neutral particle beams in SDI
        [C]   is to be used as an offensive weapon in DSI
        [D]   may be used for reconnaissance
        [E]   may be used to launch ABM's in SDI
        [F]   C and E are true
        [G]   A and D are true

[12]   Nuclear fission was first observed by
        [A]   Fermi and Wigner in USA in 1942
        [B]   Szilard and Rutherford in England in 1935
        [C]   Hahn and Strassman in Germany in 1939
        [D]   Einstein and Oppenheimer in USA in 1945

[13]   The scientist who did not play a crucial role in the development of the hydrogen
        bomb was
        [A] Andrei Sakharov [B] J. Robert Oppenheimer   [C] Edward Teller
        [D] Stanislaw Ulam   [E] they all did

[14]   In a nuclear scenario, which three of the following constitutes the strategic triad of offensive weapons?
        [A]   Warhead-bearing helicopters   [B]   long range bombers
        [C]   submarine based ballistic missiles   [D]   cruise missiles
        [E]   ICBM   [F]   ABM

[15]   The yield of the first nuclear bomb dropped over Hiroshima was equivalent to about
        [A]   one ton of TNT   [B] 10 tons of TNT   [C] 20,000 tons of TNT
        [D]   5,000,000 tons of TNT

[16]   The first self sustaining nuclear fission chain reaction was achieved in
        [A]   Germany by Hahn and Strassman in Germany in 1939
        [B]   Chicago by Fermi and Wigner in 1942
        [C]   Los Alamos by Oppenheimer and Teller in 1945
        [D]   Paris by the Curies in 1920

[17]   A cruise missile's utility lies chiefly in which of the following characteristics?
        [A]   It has indefinite range
        [B]   It flied farther than an ICBM
        [C]   It flies low enough to avoid detection by radar
        [D]   It can be recalled
        [E]   It is the only weapon that can be submarine launched [F] none of the above

[18]   The principal advantage of lasers as possible weapons is due to
        [A]   their compact size
        [B]   the speed with which their energy is transported
        [C]   the enormous energy they emit and hence their ability to destroy large cities
        [D]   their being invisible
        [E]   the cheapness of their construction
        [F]   none of the above

[19]   Limited Test Ban Treaty for nuclear weapons
        [A]   allows for underwater and underground testing
        [B]   allows for testing only in outer space
        [C]   allows for testing in the upper atmosphere
        [D]   allows for underground testing only

[20]   The country in which the largest fraction of electricity is produced by nuclear generators is
        [A]   U.S.A.   [B]   C.I.S. [C]   U.K.
        [D]   Japan   [E]   France

[21]   Which of the following places is unrelated to the others?
        [A] Chernobyl [B] Windscale [C] Monte Carlo [D] Three-Mile Island

[22]   The first USA detonation of a hydrogen bomb occurred in
        [A] 1949 [B} 1947 [C] 1952 [D] 1962

[23]   The first Soviet detonation of a hydrogen bomb occurred in
        [A] 1949 [B] 1953 [C] 1957 [D] 1971

[24] The Hungarian physicists, Leo Szilard, Edward Teller, and Eugene Wigner assembles in 1939
[A] to prepare a paper on nuclear fission for publication in a U.S. journal
[B] to inform J. Robert Oppenheimer of latest developments in fission
[C] to inform Albert Einstein of latest developments in fission and urge him to write to President Roosevelt
[D] to discuss ideas about space travel

[25] The dropping of the atomic bomb over Hiroshima in 1945 was carried out in order to
[A] demonstrate a revolutionary physical principle
[B] reduce the number of casualties in the conquest of Japan
[C] persuade the Japanese to surrender
[D] Inform the USSR of USA possession of a winning weapon
[E] justify the expenditure of two billion dollars
[F] A,D,E, [G] B,C,D [H] A,B

[26] Electromagnetic pulse (EMP) is generated by
[A] an underwater nuclear explosion
[B] the re-entry of a space shuttle into the atmosphere
[C] an underground nuclear explosion [D] a laser
[E] a high altitude nuclear explosion [F] a fast submarine

[27] The head of the Manhattan Project was
[A] Hans A. Bethe
[B] General Leslie R. Groves
[C] J. Robert Oppenheimer
[D] Edward Teller
[E] R.R. Wilson

[28] To maintain a nuclear chain reaction,
[A] at least one neutron must be emitted per fission
[B] at least two neutrons must be emitted per fission
[C] at least three neutrons must be emitted per fission
[D] at least ten neutrons must be emitted per fission
[E] no neutrons need be emitted

[29] The deployment of missiles in Cuba by USSR in 1962 may have been motivated by
[A] an attempt to deter a US invasion of the island which the Cubans believed was imminent
[B] Khrushchev's attempt to offset U.S. missiles in Turkey
[C] a Soviet pilot to cut off the supply of Cuban sugar cane to the U.S.
[D] all of the above
[E] only A and B

[30] The first artificial orbiting satellite was launched by
[A] USA in 1961 [B] USA in 1968 [C] USSR in 1957
[D] France in 1971 [E] England in 1952

[31]   Burning of carbon and hydrogen proceed by chemical reaction
       (a) $C + O_2 \rightarrow CO_2 + 4eV$ energy (b) $2H_2 + O_2 \rightarrow 2H_2O + 3eV$ energy
       [A]   We shall, in the future, need the reaction (a) as a mobile fuel
       [B]   We shall, in the future, need the reaction (b) as a mobile fuel
       [C]   Both (a) and (b) are environmentally acceptable

[32]   In general strontium and cesium are basic fission products resulting from slow
       neutron induced fission of
       [A] U_{238} [B] U_{235} [C] U_{239} [D] Plutonium (239) [E] B and D

[33]   Burning of methane is based on the reaction $CH_4 + 2O_2 \rightarrow CO_2 + 2H_2O$
       [A]   It is a process that adds CO_2 to the environment
       [B] CO_2 is absorbed by H_2O (water) and therefore not released into environment
       [C]   both of the above

[34]   In the production of energy a certain amount of waste is a necessary by product
       [A]   Waste from fossil fuel burning is localized and therefore can be controlled
       [B]   Waste from burning nuclear fuel is spread globally and cannot be controlled
       [C]   A and B are correct
       [D]   A and B are wrong

[35]   Geosynchronous orbit is at a distance
       [A]   300 miles from the earth   [B] 3 million miles from the earth
       [C]   22 thousand miles from the earth

[36]   LASER (light amplification by stimulated emission of radiation) if used as space
       weapon it can reach and destroy a missile at a distance of 2000 Km
       [A] in one minute   [B] in 1/150 second   [C] 2 seconds

[37]   In the production of nuclear warheads it is inevitable that radioactive and toxic
       material also are produced.
       [A]   Nuclear weapons complexes are isolated and therefore they do not affect
              people living in villages and cities
       [B]   Nowadays around all these plants cities are built and therefore their
              environmental impacts cannot be ignored.
       [C]   All of these weapons complexes are inherently safe and they do not pose any
              environmental threat

[38]   GPALS (Global protection against limited strikes) is the more focussed evolution
       of the SDI programs.
       [A]   It can be effectively used as a defense against possible future incoming
              ICBM's
       [B]   It is meant to be effective only for third world missiles used against any
              peaceful country
       [C]   It can be used as a United Nations defense against any irresponsible country
              intending to strike its neighbors with missiles
       [D]   B and C

[39] Of the following proposed treaties, which one (or ones) was (were) actually signed and ratified?
[A]   Limited Test Bank Treaty (LTBT)
[B]   Comprehensive Test Ban Treaty (CTBT)
[C]   Strategic Arms Limitations Treaty (SALT)
[D]   Strategic Arms Reduction Treaty (START)
[E]   Anti-Ballistic Missiles Treaty (ABMT)
[F]   Intermediate Nuclear Forces Treaty (INFT)

[40] The highest yield hydrogen weapon ever exploded had an energy equivalent to about
[A] 20 kilotons of TNT   [B] 5 megatons of TNT
[C] 60 megatons of TNT   [D] 12,000 megatons of TNT

NUCLEAR WAR-NUCLEAR PEACE ANSWER SHEET

NAME:_____   HON 331 FINAL 1992

On my honor, I have neither received nor given aid on this examination

SIGNATURE:_____   I.D.# _____

Clearly PRINT the letter or letters corresponding to your answer on this sheet. (40 questions - 2 points each)

[1]____C_____   [21]_____C_____

[2]____D_____   [22]_____C_____

[3]____C_____   [23]_____B_____

[4]____B_____   [24]_____C_____

[5]____D_____   [25]_____G_____

[6]____C_____   [26]_____E_____

[7]____E_____   [27]_____B_____

[8]____C_____   [28]_____B_____

[9]____A,F_____   [29]_____E_____

[10]____D_____   [30]_____C_____

[11]____G_____   [31]_____A_____

[12]____C_____   [32]_____E_____

[13]____E_____   [33]_____C_____

[14]____C,E_____   [34]_____D_____

[15]____C_____   [35]_____C_____

[16]____B_____   [36]_____B_____

[17]____C_____   [37]_____B_____

[18]____B_____   [38]_____D_____

[19]____D_____   [39]_____A,C,D,E,F____

[20]____E_____   [40]_____C_____

Popular Impact of the Course

The following letters are, among many others, examples of appreciation of the course, as well as the evening public lectures of the 1983-84 academic year. This was the reward, which I wish to acknowledge gratefully, of our efforts for public education on the, then, most important subject of our time.

BASCOM PALMER EYE INSTITUTE

UNIVERSITY OF MIAMI SCHOOL OF MEDICINE
DEPARTMENT OF OPHTHALMOLOGY

P.O. BOX 016880/ MIAMI, FLORIDA 33101 /U S A
TELEPHONE: (305) 547-

January 26, 1983

Professor Behram Kursunoglu
Director
Center for Theoretical Studies
341 Comp. Ctr.
University of Miami
Coral Gables, Florida 33134

My dear colleague:

Through the kind actions of your associate, Arnold Perlmutter, I received notice of your University of Miami class and the adult education lecture series entitled "Nuclear War--Nuclear Peace," and feel that I must write to you on it.

It is simply a superb gesture and, in my opinion, one of the most important lecture series that has ever been undertaken here at this University. And I say this advisedly. Certainly, you should be warmly congratulated for having the insight and the refined dignity to have arranged this at this time. As was commented the other evening at the Weisskopf lecture, Miami does not often put its best foot forward on such critical issues. But you have indeed done so, and deserve the gratitude of the whole community for this very fine effort. I shall do my best to attend the Wednesday evening sessions with some regularity, and look forward to discussions with you and your guests on these important issues.

Very truly yours,

Thorne Shipley, Ph.D.
Professor of Visual
and Cognitive Neural Science

P.S. For your information, I have asked my colleague Dr. Alan Lockwood to send you a copy of a public forum which is being planned at the Medical School on similar issues. It is to take place, as you see, on April 9, at the Knight Conference Center. I am sure that you and any of your students who wish to attend, would be most welcome.

ANNE BATES LEACH EYE HOSPITAL
900 N W 17th Street

WILLIAM L. McKNIGHT VISION RESEARCH CENTER
1638 N W 10th Avenue

PAUL D. NOVACK

8842 BYRON AVENUE
SURFSIDE. FLORIDA 33154

January 19, 1983

Dr. Behrm Kursunoglu
Director
Center for Theoretical Studies
University of Miami
Coral Gables, Florida   33124

Dear Dr. Kursunoglu,

I have written up an article on the lectures series which
has begun, and have submitted it to the Community Newspapers for publication. I expect it to run within a week.

The program which you have put together is quite impressive
and reflects well on the entire University. I hope that
each of the lectures will be well attended by members of
the community.

The Community Newspapers has 12 editions, each serving a
local community in Dade County. I hope that the article
helps inform the public about the series.

Sincerely,

Paul D. Novack

162

University of Miami
Miami, Florida 33101

DEPARTMENT OF NEUROLOGY
P.O. Box 016960   (305) 547-6731
Mailing Address: School of Medicine (D4-5)
Location: 1501 N.W. 9 Avenue

February 10, 1983

Behram N. Kursunoglu, M.D.
Center for Theoretical Studies
University of Miami
Coral Gables, FL

Dear Dr. Kursunoglu:

Congratulations on putting together a splendid series of lectures for the
members and friends of the University community. This is truly an outstanding series and is a great service to the citizens of South Florida.

The enclosed brochure describes the symposium that I thought you might be
interested in. The symposium will be held on April 9th and has been organized by physicians for social responsibility and cosponsored by the
Division of Continuing Medical Education. I hope to attend some of your
lectures. I would especially appreciate it if there were a chance to meet
Mr. Warncke at the time of his visit. He is also scheduled to speak at
our symposium, and, as I have never met him, I would appreciate the opportunity to do so.

Sincerely,

Alan H. Lockwood, M.D.
Associate Professor of
Neurology

AHL/s

2131 Dofena Lane
East Meadow, NY 11554
May 2, 1983

Dr. Behram N. Kursunoglu
Center for Theoretical Studies
University of Miami
Coral Gables, FL 33124

Dr. Kursunoglu,

As a student in your "Nuclear War-Nuclear Peace" course,
I wish to congratulate yourself and Professor Perlmutter on its
great success.
I agree that an enlightened public is essential to understanding and dealing with the nuclear problems of today. Your
course is a unique first step in informing the college community
on these difficult issues.
There can be no greater dilemma that faces the world today
than that of the possible annihilation of life on this planet.
For that reason, I strongly feel that a course such as yours
should be offered at every major university.
In my opinion, "Nuclear War-Nuclear Peace" was the most
enlightening course of my college education. Its credibility
can be attributed to the highly qualified and distinguished
speakers who presented their views. I eagerly look forward to
hearing next year's lecturers.
I also was very impressed with your last presentation in
which you told of Dr. Oppenheimer's connection with the origins
of the Center, and of your associations with Dr. Einstein and
other top scientists. The University is certainly very fortunate
to have you as a member of its faculty.
Again, my thanks for an outstanding and necessary course.
Best wishes to you for success in your current project and in
all of your future endeavors.

Sincerely,

Lawrence Dick

Lawrence Dick

University of Miami
Coral Gables, Florida 33124
VICE PRESIDENT FOR STUDENT AFFAIRS
P.O. Box 248193

April 30, 1984

Dr. Behram Kursunoglu
Director and Professor
Center for Theoretical Studies
University of Miami

Dear Behram,

   Your "Nuclear War -- Nuclear Peace" Lecture Series
presentations were a tremendous success and I commend you on
such an excellent program. I believe your program provided
an important learning experience for our community and our
University. I personally found each lecture interesting,
timely, and of vital concern to today's world-wide future.

   Again, my thanks to you for helping to provide our
University with this opportunity and congratulations on a
superb program.

                                   Cordially,

                                   William R. Butler
                                   Vice President

WRB/nlf

April 27 '83 379 The Alexandros
Dear Behram: The Perlmutter
We do appreciate The Wells
your efforts and The Maytel's
we all thank you
for these wonderful The Kakac's
lecture series. The Veriroglow's
With our love The Schuh's
and appreciation The Huerta's

064 POΔOΣ — Ακρόπολις Λίνδου
RHODES — The Acropolis of Lindos
RHODES — L'Acropolis de Lindos
RHODOS — Akropolis von Lindos
RODI — Acropoli di Lindos

PRINTED BY G. LOUCATOS - ATHENS - GREECE

The above message on behalf of the rest of the audience from the undersigned was attached to the flower arrangement on the occasion of the years' last public lecture.

166

EDUCATION FOR THE 21st CENTURY

Glenn T. Seaborg

Associate Director-at-Large
Lawrence Berkeley National Laboratory
Nuclear Science Division
One Cyclotron Road
Berkeley, CA 94720

There is ample evidence that the warning given in 1983 by the National Commission on Excellence in Education, on which I served, is as timely today as it was in 1983. For the sake of our national security and the future of our nation we must take heed. The opening lines of our report, "A Nation At Risk" were:

> Our Nation is at risk. Our once unchallenged pre-eminence in commerce, industry, science and techno-logical innovation is being overtaken by competitors throughout the world...the educational foundations of our society are presently being eroded by a rising tide of mediocrity that threatens our very future as a Nation and as a people. What was unimaginable a generation ago has begun to occur - others are matching and surpassing our educational attainments.

> If an unfriendly foreign power had attempted to impose on America the mediocre educational performance that exists today, we might well have viewed it as an act of war. As it stands, we have allowed this to happen to our-selves...We have, in effect, been committing an act of unthinking, unilateral educational disarmament.

We all recognize that we live in a rapidly changing, increasingly high technology world. I have characterized our present age as that of the Third Revolution. The Revolution of Independence gave birth to our nation and established the democratic principles on which our classical concept of "equality of opportunity" - largely through education - is based. The Industrial Revolution rewarded the American spirit of inventiveness and made us leaders in the world's economy, blessed with an

Technology for Global Economic and Environmental Survival and Prosperity
Edited by Kursunoglu *et al.*, Plenum Press, New York, 1997

167

extremely high standard of living. The Third Revolution, the Revolution of Science, has already transformed how we understand our world - through the remarkable expansion of knowledge in a few decades - and is radically altering almost every aspect of our lives. Our response to the challenges of the revolution in science will, quite simply, decide our future. Our most valuable resources are our intelligence and ingenuity. As a nation, we pride ourselves on our history of pioneering new technologies; in the future much will depend not only on that capacity for innovation but also on our general preparedness to participate in the practice and production of those technological advances. The strength of our technological and scientific enterprise will determine our economic well-being, our security and our health and safety.

Science plays a central role in the world of today. Research in basic science leads to advances in applied science and then to widespread practical applications of this acquired knowledge in the derived technology. Incremental scientific advances, as well as major discoveries, result in new technologies of great commercial importance. They can give us entire new industries, as in the case of advances in molecular biology. They can give us whole new ranges of products, as in the case of polymer chemistry. They can revolutionize other technologies and industries, as has been the case for the transistor and the laser.

Basic research leads to the creation not only of new products but also new industrial processes and manufacturing systems. These can greatly increase industrial productivity, reduce costs, and improve the quality of products. For example, advances in microelectronics are aiding the production of automobiles, steel, and many other manufactured goods. Discoveries in biology are influencing the processing and production of pharmaceuticals, foods, and chemicals.

This country cannot afford another generation of students that is ill prepared to respond to the worldwide rapid growth of scientific knowledge and technological power. The nation's future depends on them.

We must improve general science education for all of our young (and not only for those who plan to continue their education and become professional scientists, mathematicians or engineers) because we need a large number of scientifically literate, nonprofessional workers with the understanding and skills to manufacture, operate, and repair increasingly complex technological equipment. Future employment opportunities, necessary to replace jobs lost in our declining "smokestack industries," will be in areas requiring technical sophistication and will depend on a workforce endowed with a practice in learning and the flexibility of mind to adapt to a society constantly changing. The old concept of a replaceable worker standing in a production line and doing one thing over and over is obsolete. The workplace demands workers who understand the automated equipment that they use and who can adjust and repair it. They must understand and apply the statistics of quality control and make decisions which require knowledge and judgment. The definition of "basic skills" is changing to include such areas as critical thinking, problem-solving, decision-making, reasoning, teamwork, adaptability, and computer literacy.

We must actively recruit young people from what have been traditionally underrepresented populations - women and minorities. Workforce 2000 reported that "White Males, thought of only a generation ago as the mainstays of the economy, will comprise only 15% of the net additions to the labor force between 1985 and 2000."

Nationwide, in 1985 one in five 18-year-olds was African-American or Hispanic; in 2010 the ratio will be one in three. As of 1995, minority enrollment in K-12 schools was nearly 59% in California alone.

In addition to the need for trained scientists, mathematicians and engineers, and nonprofessional workers with an understanding of complex technological equipment, we need widespread understanding of science among the general population. Support of the critical need for scientific literacy is best expressed by Thomas Jefferson's famous dictum, often quoted and profoundly true:

> I know of no safe depository of the ultimate powers of society but the people themselves; and if we think them not enlightened enough to exercise their control with a wholesome discretion, the remedy is not to take it from them but to inform their discretion.

The vitality of a democracy assumes a certain "core of knowledge" shared by everyone which serves as a unifying force. I think the most horrifying example I have heard is that in one survey of adult Americans, only 45% - less than half - knew whether the Earth travels around the sun or the sun around the Earth! How basic a lack of understanding of the Universe in which we live is this! And these same people will make judgments on the issue of global warming and the commitment our nation will make to solving this potentially disastrous problem? Thirty-six percent of the adults in the same survey believed that boiling radioactive milk would make it safe! Citizens who don't know what radiation is are being asked to judge nuclear power, and citizens who know no chemistry are making decisions about toxic waste. It is fundamental to the effectiveness of our democratic system that our citizens be able to make informed judgments on the more and more complex issues of scientific and technological public policy. Decisions must be made which are of critical importance to our health and safety.

There can be no doubt that scientific literacy, a solid understanding of science and mathematics, is now more important than ever before - and there is irrefutable evidence that the skills of our youth are not only not progressing with the increasing demands, but actually are deteriorating at an alarming rate. While our nation's needs for both an educated citizenry and a technologically trained workforce have grown by leaps and bounds, our ability to satisfy those needs has diminished. We must act now to reverse the self-destructive trend.

We all have an important stake in the success of our education system, and every part of our society must be involved in meeting the challenge. Education is an investment, not an expense. The Committee for Economic Development reports that each year's class of dropouts costs the nation about $240 billion in crime, welfare, health care, and services. For every $1.00 spent on education, it costs $9.00 to provide services to dropouts. For example, about 80% of all prison inmates are school dropouts, and each inmate costs the nation about $28,000 per year.

There is good news, however. Trends in SAT scores are not universally discouraging. One positive trend is the narrowing of the gap between minority and non-minority students. Between 1989 and 1995, the scores of Mexican-American students rose 16 points on the verbal and 20 points on the math portions of the SAT. Native American student scores increased 19 points across the board. African-Americans and Asians also showed gains.

Nationwide, many proposals and programs have been developed by people committed to improving the quality of education in this country and the future prospects of our youth and of our economy.

In September 1989, governors from almost every state in the Union attended the summit conference convened by President Bush in Charlottesville, Virginia. Goal Number Four of the five National Education Goals established at that time relates directly to math and science education: "By the year 2000 U.S. students will be first in the world in science and mathematics education."

However, the problem we face is illustrated by the report, a few years ago, of the Committee on Education and Human Resources of the Federal Coordinating Council for Science, Engineering and Technology (FCCSET). This report recommends that we should increase the supply of well-trained science and mathematics teachers. Half of the newly employed teachers of mathematics, science, and English are not qualified to teach these subjects, and fewer than one-third of U.S. schools have qualified physics teachers. In part as a result of this, 30% of our high schools offer no courses in physics, 17% offer none in chemistry and 70% offer none in earth or space science.

According to the FCCSET Committee report,

In the United States today there are 2.3 million public school teachers in grades K-12. The Department of Education estimates that over the next decade, we must hire 1.6 million new teachers, or an average of 160,000 teachers a year. Yet our primary source of new teachers, college students majoring in education, has fallen 55% since 1972. Today we are graduating only about half the teachers we will need to bridge the gap in the future. If it is becoming difficult to recruit teachers, it is even harder to retain them. Twenty percent of new teachers leave during their first year, and more than half leave before the sixth year. We are currently losing thirteen mathematics and science teachers for each one entering the profession.

The task of guiding the intellectual (and often social) development of our young is an all-important one. We must begin to recognize teachers' contributions not only by adequately compensating them for their service, but also by giving them due respect which would motivate them to refine their skills and expand their knowledge to meet future challenges. There are a number of vital new programs and proposals which address this need.

There are many interesting new curriculum development projects. Among the most fascinating to me is an initiative by the American Association for the Advancement of Science (AAAS), Project 2061 (so named to make clear its goal of revolutionizing the teaching of science by the time of the arrival of the next Halley's Comet). Its first report *Science for All Americans* (1989) makes fascinating reading: it represents Phase I (Goals) of the project by attempting to define what basic core of science knowledge should be included in the education of all young Americans. Phase II (Formulation) makes recommendations on new science curricula, instructional materials, testing methods, teacher training, school organization, and educational research and development programs. Phase III (Implementation) will probably take a good deal longer to accomplish, but Project 2061 is well on its way. The project has translated the science literacy goals outlined in its first report into a curriculum design tool, published as *Benchmarks for Science Literacy*, that establishes learning goals for the ends of grades 2, 5, 8, and 12. Also in the works are a number of tools to help educators improve their curriculum, improve their own science literacy and effect change in the K-12 education system. Project 2061 plans to merge all of these tools into an interactive, multimedia tool to design curricula and serve as a resource.

I would like to conclude by emphasizing the important role that science centers can play in changing the face of education in our country. Needless to say, I will use the University of California's Lawrence Hall of Science, of which I serve as Chairman, and a description of its far-reaching programs as an example.

The Lawrence Hall of Science (LHS) is an institution committed to improving the quality of mathematics and science instruction for pre-collegiate students. For nearly three decades the LHS has dedicated its superior resources as part of the University of California to the continuing battle against educational mediocrity.

The Lawrence Hall of Science was conceived in 1958 and built in 1968 as a memorial to Ernest O. Lawrence, the University of California's first Nobel Laureate and inventor of the cyclotron. As a dynamic research and educational institution, the LHS continues today, 28 years after its dedication in 1968, to focus its efforts on three main objectives:

1. To improve the quality of mathematics and science instruction for the benefit of pre-collegiate students through the development of innovative math and science courses and accompanying curriculum materials and teacher training services;
2. To augment the mathematics and science instruction provided by our schools, offering special mathematics and science courses at the LHS; and
3. To enhance the knowledge, appreciation, and enjoyment of mathematics and science for the general public by providing the community with a math and science center.

The LHS has provided innovative leadership in pre-collegiate math and science education through the publication of major curricula. Over 20 million students and 100,000 teachers in the U.S. use LHS-produced materials. Curricula and exhibits developed by the Hall are currently used by schools and science centers in over 30 countries. Each year, educators from around the world visit the LHS to learn new techniques to improve science and mathematics instruction. For the effective utilization of these programs and materials, the Hall provides comprehensive teacher training workshops and seminars. It instructs over 20,000 teachers each year to improve their science and mathematics knowledge as well as their instruction techniques.

The LHS has about 300,000 visitors each year. It provides classes for over 75,000 students a year at the Hall and for another 150,000 children in schools throughout California.

I would like to briefly describe a few of our many programs.

Curriculum Development:

• Chemical Education Material Study (CHEM Study) is an integrated program of written materials and films developed to improve the teaching of chemistry at high school level. It emphasizes the discovery method and the experimental approach to learning. Materials from this program have been translated into 17 foreign languages, and are used throughout the world.

• Great Explorations in Math and Science (GEMS) is a curriculum development program originally funded by the Carnegie Corporation and the Mellon Foundation to document and disseminate the Hall's well-tested science and mathematics instructional activities and lesson plans. GEMS also includes some of our exhibits and their accompanying learning activities so these may now be used by other science centers and museums. To date, 50 GEMS guides have been published, including teacher and leader handbooks, exhibit guides, assembly guides, and a parent's guide.

• Science Curriculum Improvement Study (SCIS) is a course content improvement project that has developed a sequential, articulated elementary school science program. This program is based upon the structure of science as seen by contemporary scientists, is consistent with a current view of the intellectual development of children, and reflects the experience of elementary teachers working with preliminary SCIS units. Developed in the late 1960's, it is now used in over 20% of the nation's elementary schools.

• Full Option Science Systems (FOSS) is a complete, activity-based curriculum for elementary school science instruction, grades K-6. Twenty-seven modules, covering earth, life, and physical science, guide teachers and students through creative, challenging, real-world science investigations. Each FOSS module is a complete classroom resource, including a kit of student materials, extensive teacher guide, and teacher preparation video.

• The Science Education for Public Understanding Program (SEPUP) designs instructional materials - for elementary and secondary schools and the community - that focus on science and technology and their interaction with people and the environment. These learner-centered materials promote the use of scientific principles, processes, and evidence in public decision making and are appropriate for students of all ethnic, cultural and socio-economic backgrounds, especially students at risk. The SEPUP approach also enhances the role of teachers as facilitators of student learning and as educational leaders within their communities by having them share in the development, implementation and assessment of issue-oriented science materials and programs.

Teacher Training and School Programs:

• The Center for School Change. The Lawrence Hall of Science Center for School Change helps schools and school systems to respond effectively to the rapid pace of change in our society. The Center provides sustained, coordinated support for K-12 schools interested in achieving long-lasting systemic reform. Such support enables schools to develop comprehensive, inquiry-based science and mathematics programs. The Center focuses on the following elements of systemic reform: Curriculum Coordination,

Integration and Alignment of Standards; Coordinated Professional Development; Equity and Access; Performance-Based Assessment; Parent and Family Involvement; Educational Technology; and Pre-Service Education.

• EQUALS encompasses programs for teachers, counselors, administrators, and parents. It promotes participation of students and adults, particularly women and girls, in mathematics courses and encourages their interest and involvement in math-based fields of study and work. LHS provides teacher training through programs such as EQUALS to over 10,000 teachers each year.

• Family Math teaches parents how to help their children with math at home, informs parents of the role mathematics plays in their children's studies and career choices, and creates a family enjoyment of mathematics. The project provides training and curriculum materials to help parents and teachers establish Family Math classes in their schools.

Student programs:

• ACCESS is an effort of UC Berkeley and the Oakland School District to strengthen the capacity of the District's secondary schools to prepare minority students for college. University staff work daily at school sites providing a wide range of technical assistance, training, and student support to improve mathematics, English, and science curriculum and instruction; school management policy and organization; counseling; and parent involvement.

• Center for Multisensory Learning (CML) specializes in science and computer education for disabled students. CML staff conduct workshops for disabled students at the Hall, train teachers to use the SAVI/SELPH science program in their classrooms, and organize special events that enhance opportunities for disabled students to enjoy science and technology experiences.

• The Public Science Center. The Lawrence Hall of Science Public Science Center endeavors to reach the widest possible audience in the museum setting and beyond. Using the Hall's innovative methods and materials, LHS creates effective exhibits, programs, and activities that show how science and mathematics concepts can produce enjoyable, accessible, and relevant educational experiences for families and schools. The science center is filled with experiences that make science and mathematics fun and personal, including engaging exhibits that immerse visitors of all ages in hands-on explorations; a biology lab of live animals to see and hold; the LHS planetarium, where the audience members become scientists; stargazing from the LHS plaza; and playful scientific sculptures that allow for learning and climbing, while enjoying a spectacular view of the entire San Francisco Bay Area.

We must work together, employing our considerable resources, to ensure a prosperous future. Whatever the expense of improving science and mathematics education, this is an investment in the future we must make. Excellence is costly. But in the long run, mediocrity costs far more.

CHAPTER IV
GLOBAL INITIATIVES FOR THE FUTURE OF NUCLEAR ENERGY

THE LOS ALAMOS NATIONAL LABORATORY NUCLEAR VISION PROJECT

Edward D. Arthur and Richard L. Wagner, Jr.

Los Alamos National Laboratory
Los Alamos, New Mexico USA

ABSTRACT

Los Alamos National Laboratory has initiated a project to examine possible futures associated with the global nuclear enterprise over the course of the next fifty years. All major components are included in this study—weapons, nonproliferation, nuclear power, nuclear materials, and institutional and public factors. To examine key issues, the project has been organised around three main activity areas—workshops, research and analyses, and development of linkages with other synergistic world efforts. This paper describes the effort—its current and planned activities—as well as provides discussion of project perspectives on nuclear weapons, nonproliferation, nuclear energy, and nuclear materials focus areas.

INTRODUCTION

Dramatic global political changes, including the end of the Cold War, have provided both an opportunity and incentive for an examination of the future of the global nuclear enterprise. The Nuclear Vision Project at Los Alamos National Laboratory is intended to explore how the evolution of "things nuclear" might progress globally over the next half century. Essential elements of this study include the future of nuclear energy and other civil applications; nuclear weapons and proliferation; the institutions associated with, and public opinions about, nuclear science and technology; and environmental and other related issues. The general objective of the project is the creation of a "global vision" which seeks to manage the balance between realising nuclear technology benefits globally and reducing nuclear dangers.

The efforts described here occur in an environment of past similar efforts, current efforts underway at institutions such as other Department of Energy laboratories, or efforts planned to begin shortly. As examples, these include the U.S. National Academy of Sciences study on excess weapons plutonium[1], the American Nuclear Society study on protection and management of plutonium[2], the International Nuclear Societies Council vision for the second fifty years of nuclear energy[3], the Sandia National Laboratories

Technology for Global Economic and Environmental Survival and Prosperity
Edited by Kursunoglu *et al.*, Plenum Press, New York, 1997

project on the role of American nuclear weapons in the first quarter of the twenty-first century[4], The U.S. Nuclear Posture Review, the Atlantic Council study on the future of nuclear power in Asia and globally[5], and the proposal[6] for an international commission to assess the future of nuclear power globally.

GOALS

The Nuclear Vision Project does not attempt to predict the future but rather aims to explore alternative futures and their implications. In doing so, we hope to gain a deeper understanding that will guide actions at Los Alamos, where the great majority of research and development occurs in areas directly linked to nuclear technology. In addition, we would like project results to provide credible information relevant to current and future policy dialogues on major nuclear issues.

The project seeks to examine alternative futures in the context of important realities that carry implications for the future: the increasing demand for nuclear energy in east Asia, the growing global inventories of nuclear materials, an increasing demand for energy coupled with the finiteness of many current fuel sources, the reduction of superpower stockpiles, the increasingly rapid diffusion of nuclear technical knowledge and capabilities across the globe, and the risks of nuclear weapons proliferation and of proliferation of other weapons of mass destruction. Exactly where these current situations may lead is unknown, but it is certain that the future will be shaped both by unanticipated events and intentional actions. The task of this project is to think broadly about what is desirable and possible in the context of some set of future realities, without departing from the realm of possibility.

METHODOLOGY AND APPROACH

Our attempts to examine a "global vision" are conditioned by the practical need to tie long-range objectives back to near-term actions and decisions through achievable steps and milestones. The process starts with the positing of desirable futures (endstates) 20 to 50 years into the future. The decision to address a nominal 50-year horizon represents a realistic boundary given that technology development and implementation generally take decades. Some in the energy industry plan at least two or three decades ahead, and individual military systems, including ships, aircraft, and nuclear weapons, can remain in inventory for several decades. Such posited futures are intended to include the major nuclear-related elements as well as to explore relationships among them. Linkages and impacts on outside factors associated with areas such as global stability and security or global, regional, and national energy security are under investigation as well. Finally, these posited futures will be tested against some series of scenarios that could impact them positively or negatively.

Once a set of nuclear futures has been proposed, the next step in the process involves identifying key technology, institutional interface, and policy steps that should occur along the way. Since expertise at Los Alamos National Laboratory lies primarily in forefront areas of technology development, identification of needed science and technology is a component of major interest to the project. The final step involves taking information developed as a result of the first steps to identify actions in technology, institutional, and policy areas that would allow one to proceed along a "roadmap" towards a desired nuclear future, or some major component of such a future. Examples of such a process are provided later in the paper.

To approach these complex nuclear-related areas, the project has been divided into three elements—internal and external workshops, internal Los Alamos research and analysis projects, and efforts to link work at Los Alamos with synergistic efforts throughout the United States and the world. Figure 1 illustrates the connectivity of these three essential elements. Development of both understanding and (hopefully) consensus on complex and interrelated global nuclear issues are intended products of the project, and its success will be measured by how and whether the fruits of the project impact actions and decisions.

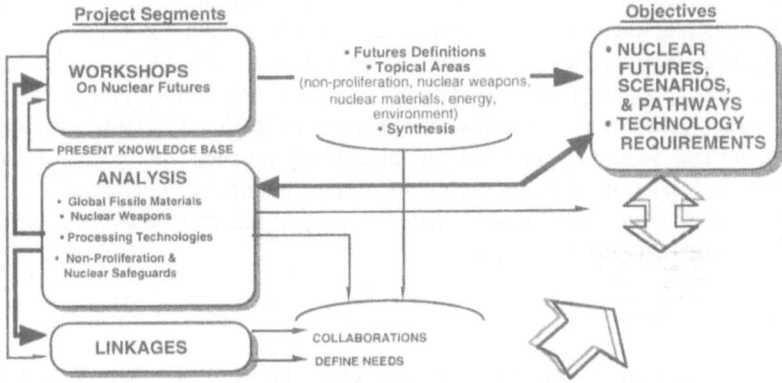

Figure 1. Los Alamos Nuclear Vision Project elements. The workshop component's duration will be approximately 1 to 1 1/2 years; the analysis efforts are intended to last for approximately 3 years; and the efforts to establish linkages with synergistic efforts will be ongoing.

Workshops

Three workshops have been held on the topic areas outlined earlier. Each workshop was designed to include experts who contribute a broad spectrum of perspectives and opinions, ranging from general support of the benefits of nuclear technology to legitimate concerns about present and future nuclear activities.

The first workshop (August 1995) was entitled "Securing the Nuclear Future." Its objective was to launch the Nuclear Vision Project by discussing the present status and long-term implications of the global nuclear endeavour. The title was chosen to reflect the theme of searching for desirable nuclear futures that achieve the project objectives outlined earlier and which deal with current and expected nuclear realities.

A second workshop on "Nonproliferation and International Security" (November 1995) consisted of presentations and discussion on three major topics aimed at assessing the nuclear future in the 2015 time frame from nonproliferation and international security perspectives. These topics were a) the bases and implications associated with linking arms control and nonproliferation; b) material connections involving plutonium, future situations in the former Soviet Union, and nuclear smuggling; and c) research and development priorities in nuclear material, protection, control, and accountability areas.

The third workshop on "Nuclear Weapons and Stewardship Issues" (April 1996) used the concept of regimes (defined to mean collections of agreements, treaties, and capabilities that describe a global norm) as a major thread for discussion. This workshop examined the conditions, features, and issues associated with past, present, and future nuclear weapons regimes. The roles of nuclear weapons in the future, associated dangers posed by nuclear arsenals, and the challenges in defining desirable regimes for the future were examples of topics discussed at this workshop.

Future workshop plans include one on global nuclear energy and nuclear materials futures to be held later in 1997, tentatively entitled "Nuclear Energy and Global Security".

Research and Analyses

To complement and support activities associated with workshops and efforts to establish links with other activities, a small internal effort exists to examine key issues. Areas under investigation include utilization of models to examine future global and regional energy scenarios with emphasis on how nuclear power competes with other energy sources on the basis of drivers such as economics, environmental requirements, and proliferation drivers. A particular focus of this effort is understanding future dynamics of nuclear materials (growth of plutonium contained in spent fuel stocks, for example) and the development and application of quantitative methods for assessing proliferation risks associated with nuclear materials sources. In areas pertaining to future environments surrounding nuclear weapons, analyses are also being pursued that examine stability relationships and nuclear weapons roles in future multipolar international security environments. Discussion that follows in latter sections of this paper elaborates on current and future investigation activities.

Links to Related Efforts

A great deal of work relevant to the Nuclear Vision Project is being conducted outside of Los Alamos. By developing collaborations with appropriate individuals and groups throughout the U.S. and internationally, the project can establish important linkages to other institutions and centres of expertise associated with future nuclear issues and systems.

DISCUSSION

Where do we stand in our appreciation of the nuclear future? To oversimplify somewhat, we will divide the discussion into two parts: 1) global nuclear energy and 2) nuclear weapons and (non)proliferation. To the extent that these topics are linked, it is mainly by the global nuclear materials posture—civil and military—and possible approaches to managing it. Part of our approach is to study the nuclear materials issues as they pertain to energy and to weapons and then to integrate them later in the project. Figure 2 provides an overview of research activities underway or planned for the project and illustrates connections between them as well as to global security and energy security externalities.

Nuclear Energy and Nuclear Materials

Our approach to examination of nuclear energy scenarios focuses on 1) understanding drivers that can impact future trends in global, regional, or national use of nuclear power, and 2) issues associated with nuclear power and technology/fuel cycle/institutional approaches that deal with them. Both of these two areas work towards definition of a desired nuclear power "system" for the 2050 time frame—one that has strengthened attributes in areas such as economics, safety, and material management—and the steps that can lead up to it.

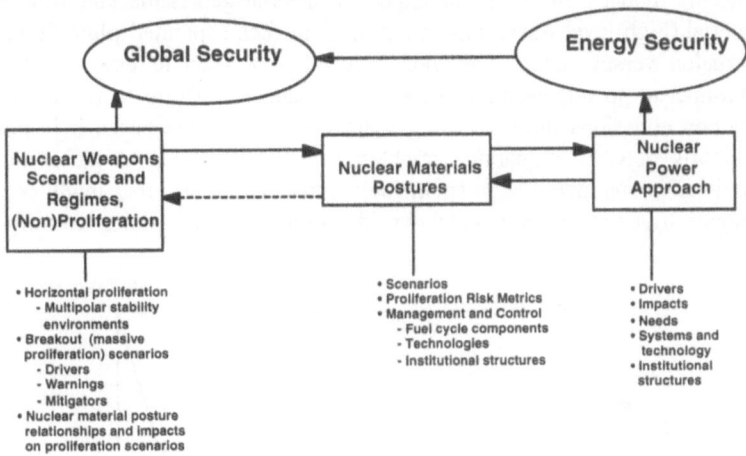

Figure 2. Analysis focus areas of the Nuclear Vision Project

Key questions that guide our examinations include:

- What are the linkages and tradeoffs associated with future global energy security—including meeting energy and economic growth needs of developing nations—and nuclear power? How does the need to counter negative environmental factors such as emissions from fossil fuels impact such relationships?

- What is the role of plutonium in future global nuclear energy scenarios? What technologies can prove effective in managing plutonium—both to utilize its energy content and minimize inventories that exist in a range of forms during the fuel cycle and which eventually must be disposed of in long-term geologic storage?

- What is the nature of the relationship between the growing global inventory of nuclear materials from civilian nuclear power and possible future dangers associated with proliferation of nuclear weapons? (See discussion in the following section on Nuclear Weapons and (Non)Proliferation.) Can proliferation risks from such material sources be quantified credibly so as to allow meaningful comparisons with other routes for material production? Are current safeguards for nuclear materials sufficient or are extensions needed? Could an inventory reduction strategy be applied? What technologies and institutional means would be attractive?

To provide an initial basis to address these topics, we utilize the Edmonds, Reilly, Barnes global energy, economics, environmental (E^3) model[7] modified to provide more detail in its treatment of the nuclear fuel cycle, particularly in the areas of reactor/fuel cycle economics and nuclear materials flow[8]. This framework allows a consistent set of projections for the amount (as competed on the basis of economics, resource availability, and projected energy needs) of nuclear power that could be implemented in each of nine major world regions, as well as globally. Figure 3 illustrates some initial (unbenchmarked) nuclear power and plutonium inventory scenario results[8] from this model. As indicated by these results, the model also computes expected material generation and flow amounts, types of material (high-level waste, plutonium in spent fuel, separated plutonium, etc.) for each major region versus time. The model can also be used to examine other future impactors on nuclear power, such as (economic) impacts of efforts to reduce fossil-fuel-produced carbon emissions through taxing structures. With this basis, it is possible to examine tradeoffs between disparate variables, such as reduction of global or regional carbon emissions versus increases in spent-fuel source plutonium, that might occur under energy strategies driven by environmental considerations (acid rain, global warming, etc.).

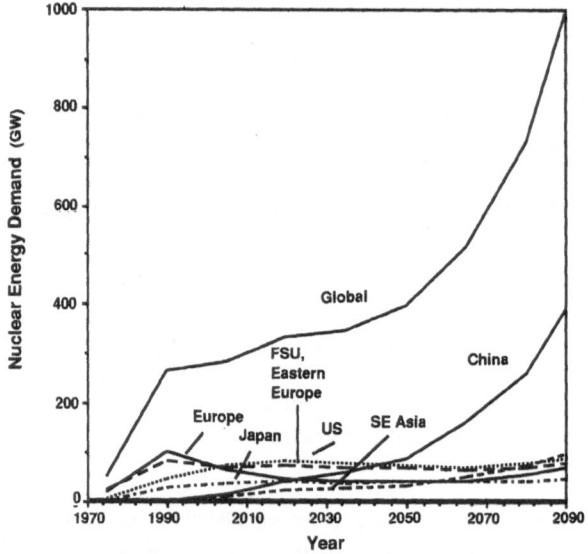

Figure 3a. Model results for projected nuclear power demand (gigawatts of electric power).

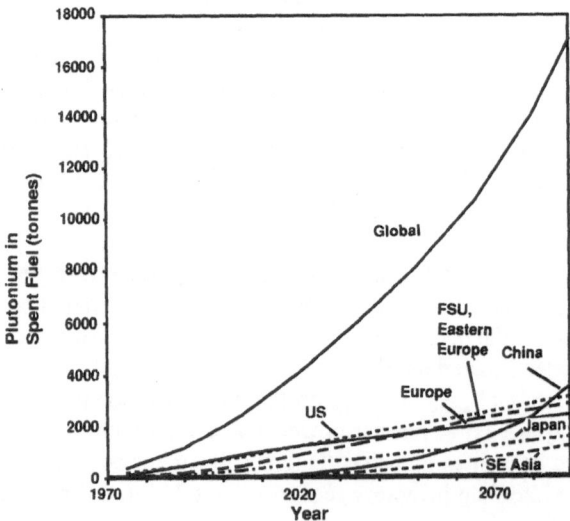

Figure 3b. Model results for projected plutonium inventories in reactor spent fuel (assuming a once through fuel cycle).

Results from such a model-based approach will be augmented with efforts involving two other areas. The first area deals with the assessment and grouping (by broad categories) of current and future national and/or regional situations regarding energy security. This effort is a "top-level" attempt (using much information obtained via the efforts to link the Los Alamos Project with other synergistic efforts) to better understand how nuclear power may fit into energy security strategies, particularly those associated with Asian nations' development.

The second area of effort involves extension of past analytic work[9,10] aimed at the quantification of proliferation risks associated with continued world-wide deployment of nuclear power. The focus of this effort is examination, from proliferation perspectives, of global stocks of plutonium-containing spent fuel to assess risks against other material production/diversion routes that may be used for proliferation or massive nuclear weapon stockpile creation or reconstitution (see Nuclear Weapons and (Non)Proliferation discussion that follows). Features of interest for evaluating such weapons-related aspects of a global nuclear materials posture include:

- Total quantities of weapon-usable material—nationally, regionally, and globally— in various stages of a range of nuclear fuel cycles

- The chemical and isotopic forms of the material

- Future technologies for storage, security and monitoring, transportation, and separations

- Material location and how (physically and institutionally) it is secured and monitored.

This effort will also examine ways to address quantitatively long-term risk (in areas such as safeguards requirements and toxicity impacts) that are associated with the disposal of large amounts of spent fuel containing plutonium. For example, plans are underway to evaluate technologies and scenarios for recovery of plutonium in spent fuel after it has been placed in geologic disposal.

These tools can be used to assess several plutonium management strategies—a once-through cycle followed by geologic disposal, a plutonium recycle system possibly followed by plutonium breeding, or other fuel cycles such as thorium-uranium—from perspectives of short- and long-term risks arising from nuclear materials. We have already developed expanded models to predict growth in future plutonium inventories and to assess the impact of technologies for dealing with them. These studies[11] are also examining strategies and approaches for situating and managing operations within the nuclear fuel cycle. In summary, we are building a suite of tools that will allow us to evaluate alternative future nuclear materials postures in the context of a range of possible future regional and global security environments.

A desired end result from these efforts, as well as those developed in working relationships with other interested parties, is creation of a clearer picture of, and perhaps even a high-level roadmap towards, a desired nuclear power system endstate that would exist in the year 2050. Such a picture would address fuel cycle options and strategies; institutions that might exist or be created to enhance nuclear power in areas such as safety, material management, safeguards, research and development, etc.; and components and technology associated with nuclear power in 2050. An example of one part of such a picture and the process leading to it is illustrated in Figure 4 where a progression of activities and technologies are identified for the management of plutonium from excess weapons and spent fuel.

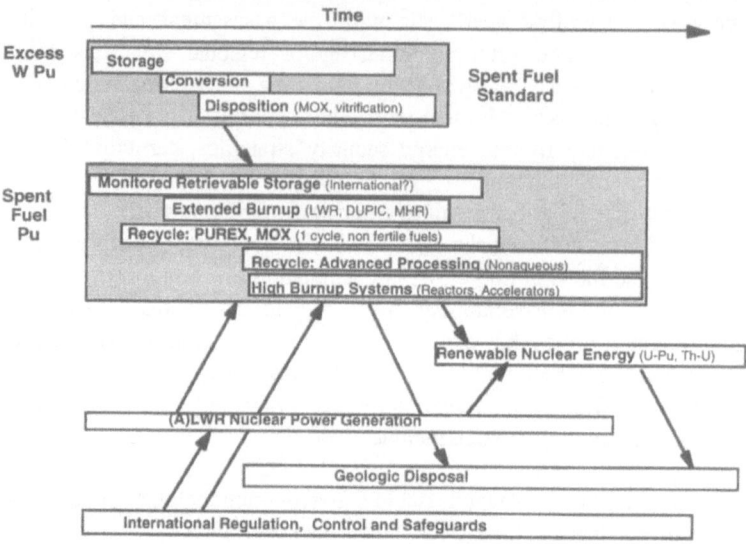

Figure 4. "Roadmap" example for strategies and technologies applicable to plutonium management.

Nuclear Weapons and (Non)Proliferation

Two main themes organise our approach to thinking about nuclear weapons and (non)proliferation: 1) understanding the dangers associated with nuclear weapons over the long term and 2) understanding possible evolution of congruent behaviours—negotiated or tacit, unilateral, bilateral, or multilateral—on the part of nations/polities by which the dangers can be contained.

Containing future nuclear weapon dangers has many facets. For the purpose of this paper. we can group future nuclear weapon-related scenarios into two broad categories:

- "Canonical" horizontal proliferation of the general type represented by North Korea, Iraq, India/Pakistan, etc.

- Possible (farther) future cases in which the world situation might develop in such a way that many nations/polities/blocs see a need (or an increased need) for nuclear weapons. Such situations could lead to possibly widespread, possibly rapid, competitive, horizontal and vertical, nuclear (re)armament ("breakout").

Canonical horizontal proliferation (the first category) is much thought and worried about and is, in a sense, the baseline case. In the project, we extend examination of this category to analyse instabilities (and therefore nuclear dangers) arising from two principal trends: 1) erosion of the classic conditions of deterrence (a bi-polar environment, U.S. versus Soviet Union) to include multipolar interactions and 2) reduction of numbers of nuclear weapons possessed by current nuclear weapon states coupled with possible expansion of the numbers of·countries holding nuclear weapons so that possession of a "few" is amplified in its relative importance. We are also focusing attention on the "breakout" case for several reasons. First, it is simply less well understood. (It is also less understandable, because it would presumably develop, if at all, in the farther future, and because the only historical case—the early decades of the Cold War—is probably a poor model.) But more importantly, we feel that, in light of the destructiveness of nuclear weapons, prudence demands looking at such limit cases.

Attention to large scale (re)armament scenarios is suggested also because, as the global technology base grows and spreads over our long time horizon, the technology, science, and industrial capabilities needed for nuclear weapons development and production become more widespread. Capabilities to create substantial numbers of nuclear weapons (perhaps quickly) would thus be "latent" in many more nations' scientific, technological, and industrial bases (as they were in a few nations in 1942). One might say that these nations will "have" "latent nuclear weapons."[a]

In reality, there is a continuum in a conceptual sense between the extremes represented by our two categories of scenarios. They could be connected over time as well; an increasing rate of canonical, horizontal proliferation could trigger a nuclear "breakout" as posited in our second category. Thus, we are conscious of the need not to give short shrift to nearer-term, "smaller," canonical proliferation—in part because of connections with larger scenarios, but also because it is important in itself.

However, the reverse—neglect of large-scale nuclear (re)armament scenarios—is too often the case in the current conventional wisdom. The emphasis on *preventing* proliferation (deeply embedded in the current, Nonproliferation Treaty-oriented regime) diverts attention from developing ways of *managing its consequence* if it occurs. Also, thinking about how to deal with a canonical proliferation crisis (including, possibly, a "next use" of nuclear weapons) often focuses almost entirely on the immediate outcome, with little attention given to longer-term consequences involving the full range of possible nuclear dangers.

With these considerations in mind, we are beginning to try to understand the dynamics of future nuclear weapon proliferation ranging from expansion of the numbers of nations having small numbers of nuclear weapons to the other bound of widespread, rapid, large-scale, n-sided, competitive nuclear (re)armament. Among other things, we are extending models used during the Cold War to analyse two-sided crisis-stability related to nuclear (and conventional) *forces-in-being*, to try to understand the much more complex scenarios involving *latent* nuclear forces and more parties.

We also want to understand more about the technological dimensions of future arms races as well, and it is here that connections with the global nuclear materials posture may begin to enter. In particular, what would be the pacing technological/industrial factors in future nuclear proliferation and possible (re)armament scenarios? Candidates are: availability of nuclear material; nuclear warhead design and fabrication (given nuclear material); and the nuclear weapons *systems* (delivery platforms, command and control, etc.). Although it demands a much more careful look, general considerations seem to indicate that the latter two capabilities may exist widely in a more technologically/industrially advanced future. So, it will behove us to understand in some detail how alternative future nuclear material postures might affect nuclear material availability for weapon purposes in the scenarios we have discussed here.

A principal consideration is whether nations would construct dedicated military facilities or divert/access fissile material from the civil nuclear power fuel cycle. This national decision would be the result of a complex set of drivers including national and international incentives and disincentives, as well as a number of factors associated with routes that could be chosen for material production. These factors include items such as technical implementation, complexity, cost, detectability, capacity, and quality of material. Examples from history are of interest, but situations may be very different in the future. For example, separations technologies may be easier, but monitoring/surveillance may be better. In general, the more urgent the felt need for material, the more weight that could be put on access to (diversion from) the civil fuel cycle, either in terms of recovery of material

---

[a] The terms "virtual" nuclear weapons or "virtual proliferation" are sometimes used in somewhat the same sense. We prefer "latent," as it connotes time evolution—moving from latent weapons to real ones—which is of the essence of these scenarios.

from spent fuel or from reactor fuels themselves. Thus, figures of merit for evaluating alternative future approaches (dedicated reactors, clandestine separations facilities, uranium enrichment, civilian reactors and reprocessing facilities, alternative fuel cycles) should weigh heavily time to acquire materials and fabricate weapons under a variety of nuclear materials postures. The discussion on nuclear power and nuclear materials has highlighted our intended approach to this problem.

Of particular interest is the extent to which total quantities of potentially accessible material make a difference to such proliferation and breakout scenarios. To what extent, for example, is total quantity related to the efficacy of securing materials? This is one question, of many, related to what we believe are the two basic alternative strategies for nuclear material futures:

- Continued accumulation of spent fuel and other fissile materials requiring long-term security and safeguards

- Development and implementation of (possibly new) reactor technologies and fuel cycles such that, over many decades, total/regional quantities of fissile material of concern in the situations discussed above can be minimised (through long-term holdup in nuclear system cores, through reuse, and eventually through burndown of materials).

Implicit in the foregoing discussion is the idea that future global nuclear materials posture(s) might be more deliberately managed. As part of our project, we are beginning to explore how and to what extent a *concert of interests* (governments, multi-national and international political entities [already existing or new], industries, non-governmental organisations, etc.) might develop, or be developed, to this end.

CONCLUSION

March 17, 1996, was the 50th anniversary of the Acheson-Lilienthal Report, which began efforts to place all applications of nuclear science and technology under international control. The implications for mankind of the potential—for good and ill—of the energy of the nucleus of the atom are still global, indeed more so today than in 1946. The fact that this potential will continue to exist in future world affairs is a key driver for the Los Alamos Nuclear Vision Project.

REFERENCES

1. J. Holdren (Chair), Committee on International Security and Arms Control, *Management and Disposition of Excess Weapons Plutonium*, National Academy Press, Washington, D. C., 1994.

2. G. Seaborg (Chair), American Nuclear Society Special Panel Report, *Protection and Management of Plutonium*, American Nuclear Society Publication, 1995.

3. M. Hori (Chair), International Nuclear Societies Council Fifty-Year Vision Committee, *A Vision for the Second Fifty Years of Nuclear Energy*, American Nuclear Society Publication, 1996.

4. R. Rinne, Sandia National Laboratories, USA, personal communication, 1996.

5. D. L. Guertin, The Atlantic Council of the United States, proposal for a study on "The Long Term Future for Nuclear Energy", personal communication, 1996.

6. P. Beck, as proposed in *Prospects and Strategies for Nuclear Power,* Earthscan Publications, London, 1994.

7. J. Edmonds and J. M. Reilly, *Global Energy: Assessing the Future*, Oxford University Press, New York, 1985.

8. R. A. Krakowski, "Global Energy Modeling in Support of Understanding Long-Term Nuclear (Materials) Futures (Work in Progress)", Los Alamos National Laboratory document, LA-UR-96-1931, June 1996.

9. See for example, C. D. Heising, "Quantification of Nuclear Diversion Risks: Promises and Problems", Energy Policy 10, 101 (1982).

10. P. Silvennoinen, *Nuclear Fuel Cycle Optimization: Methods and Modeling Techniques*, Pergamon Press, Oxford (1982).

11. J. W. Davidson, R. A. Krakowski, E. D. Arthur, "Impact of Accelerator Driven Transmutation Concepts on the Management of Global Plutonium Inventories", Los Alamos National Laboratory document, LA-UR-96-1868, May 1996.

UNITED STATES NUCLEAR TECHNOLOGY -
NEED FOR A NEW APPROACH

Clinton Bastin

Chemical Engineer, United States Department of Energy[1]
Immediate Past President, National Treasury Employees Union Chapter 228[2]
Signatory and a Director of the Eagle Alliance[3]
14355 Long Channel Drive, Germantown, Maryland 20874

GREAT BENEFITS OF NUCLEAR TECHNOLOGY -
POTENTIAL FOR MORE BENEFIT

Nuclear technology provides great benefit to mankind and has potential for much greater benefit. It is essential for best medical diagnosis and treatment; it is essential for exploration of deep space, national and global security, and many industrial applications. The growing awareness of environmental dangers from over reliance on combustion of fossil fuels; the expanding populations and growing prosperities in many nations; the strategic importance, versatility, and diminishing supplies of petro-fuels; the limitations for cost-effective use of solar and geothermal energy; and the diminishing potential for use of fusion energy, will lead to the recognition that efficient and increased use of nuclear technology is essential if we are to provide energy for future global needs at reasonable costs and in a healthy environment.

Through proper management and efficient use of nuclear materials, instead of their disposal, long term energy needs can be met with minimum adverse impact on our environment, the full potential for benefit from nuclear technology can be achieved, and potential for misuse of weapon usable materials for destructive purposes can be greatly reduced or eliminated.

---

[1]The opinions expressed in this paper are those of the author and do not necessarily reflect those of the United States Department of Energy or the National Treasury Employees Union.

[2]The National Treasury Employees Union represents Department of Energy Headquarters Employees.

[3]The Eagle Alliance is a not-for-profit organization whose mission is to provide accurate information to America's citizens and leaders about the great benefits and potential benefits of nuclear energy, science and technology, and ideas on how this technology can be better managed to minimize potential for danger and avoid wasteful expenditures of money.

Technology for Global Economic and Environmental Survival and Prosperity
Edited by Kursunoglu *et al.*, Plenum Press, New York, 1997

189

PROBLEMS

Like many technologies of modern - and ancient civilizations and nature, nuclear technology has potential for danger and destruction. The use of nuclear technology in the United States has been and is one of mankind's safest-ever undertakings. It is important that the systems for assurance of safety and environmental protection, cost-effective management, and development of policies for use of nuclear technology be well conceived and of high quality. In the United States they are not. Because they are not, many billions of dollars are wasted, work that should be performed is not performed, nuclear technology has lost credibility with citizens of America and their leaders, and there is potential for loss of many of the great benefits. It is well past time for a new approach.

NEED: PARTNERSHIP INTERACTIONS BETWEEN LABOR AND MANAGEMENT, AND BETWEEN REGULATORS AND THOSE REGULATED

Shortly after the accident at the Chornobyl Nuclear Power Plant, I was a participant in a conference for improved relations between the United States and the Soviet Union. Most participants were not familiar with nuclear power plant operations, so I volunteered to make a presentation about this accident.

As I discussed the positive reactivity coefficient with loss of coolant, and the presence of moderator at the ends of control rods that would initially increase reactor power as control rods were inserted in an attempt to shut down operation, I noticed some individuals that seemed particularly interested in my remarks. When I finished, they asked to speak with me, identified themselves as engineers that had worked at the Kurchatov Institute in Moscow on design of nuclear reactors, then said that they were part of a large group of engineers and scientists that had protested the design of the Chornobyl type reactors - to managers, and then to senior officials of the Soviet Union. Their protests were not only rejected, they were fired. Managers made decisions that not only did not incorporate, but totally rejected best knowledge and punished those that came forward with that knowledge. Management decisions without best knowledge from employees that have resulted in catastrophes are not unique to any nation.

The explosion and loss of lives of astronauts aboard the United States Space Shuttle Challenger were the direct result of such a decision. Engineers working on the U.S. space shuttle program knew about problems of O-Rings at connections of segments of the solid rocket booster - and knew that cold weather conditions would increase the danger. If there had been an active partnership between workers and managers in the U.S. National Aeronautics and Space Administration (NASA), the knowledge from workers about the dangers would have forced a decision that the shuttle not be launched. Christa McAuliffe would still be teaching school.

In April 1996 the Central Committee of Russian Trade Unions of Nuclear Power and Industry Workers sponsored an international conference on nuclear power plant safety in Moscow to commemorate the tenth anniversary of the Chornobyl accident. My keynote address at this conference featured experiences of engineers at NASA and the Kurchatov Institute as examples of the importance of labor-management partnerships to safe operation of facilities that use complex technology and involve potentially hazardous operations or materials.

There was one other American at this conference, Ron Owen, Deputy Director of the London Coordinating Center of the World Association of Nuclear Operators, an outgrowth of the U.S. Institute for Nuclear Power Operations. Ron and I had several discussions about the concept of labor-management partnerships for ensuring safe operations of nuclear power plants. He initially challenged my remarks, pointing out the existence in U.S. nuclear programs for "employee concerns" and "whistle-blower protection" that would provide for input from employees. I countered with the argument that the U.S. concepts were inherently negative, and that input would almost always be after decisions had been made and thus likely to be rejected. Had "whistle-blower protection" been available for engineers and scientists at the Kurchatov Institute, they would not have been fired - but it is not likely that design changes would have been made in the Chornobyl type reactors. And there was probably little concern for loss of jobs by engineers knowledgeable of problems with the Challenger Shuttle launch; their concern was for lives of the Astronauts - but fell on deaf ears.

Replacement of the existing provisions for "employee concerns" and "whistle-blower protection" by a requirement for full partnerships between labor and management in all U.S. nuclear programs and nuclear activities would enhance safety, environmental and radiation protection, safeguards, cost effectiveness - all of the characteristics that contribute to a quality type operation.

Similarly, replacement of the confrontational regulatory approach of both the Nuclear Regulatory Commission for commercial nuclear programs, and the Defense Nuclear Facilities Safety Board for Department of Energy nuclear activities, by efforts to listen to and seek to understand each other, and resolution of differences through partnership type processes, would also lead to safer and more viable and cost effective nuclear activities.

The concept of partnership between workers and managers is not new, and is applicable not only to safety but to quality and cost effectiveness of all activities. Partnership agreements provide that input from workers, who have the detailed knowledge of all activities, be incorporated in decision processes - at the outset, and as the project or program moves forward. The concept has been practiced in nuclear power development programs in Japan for many years - and I learned at the conference in Moscow that it is standard practice in other nations. The practice is beginning in the United States, but it is far from standard - or even typical.

National Treasury Employees Union President Robert M. Tobias has for many years held a vision for major improvement in U.S. Government operations through partnerships between workers and managers. Ten years ago he presented these ideas to Treasury Department executives, and labor management partnerships were formed to improve quality of operations. Where there has been a major commitment by management and workers for a partnership, there has been great improvement in quality of operations.

Vice-president Al Gore has long held strong ideas of the importance of safe operation and good environmental practices for industrial facilities, and he also has strong interest in improved operation of our Federal Government. Soon after his election as Vice President, he met with Mr. Tobias to obtain an understanding of his ideas of labor-management partnerships as a means to improve quality of Government. These ideas were incorporated in Executive Order 12871, issued by President Bill Clinton on October 1, 1993.

Several weeks before President Clinton signed this Executive Order, our Secretary of Energy Hazel O'Leary met with Mr. Tobias and me and asked that we work with the Department's managers through partnerships to build needed quality in DOE operations.

DOE's Office of Nuclear Energy, Science and Technology (NE), where I work, has responsibility for ensuring availability of isotopes for health care, research, and industry; supply of radioisotope thermoelectric generators (RTGs) for exploration of deep space; development of advanced nuclear reactor and nuclear fuel cycle concepts; ensuring the safety of DOE reactors; enhancing safety of Soviet design reactors, and assisting in the stabilization of the damaged reactor at the Chornobyl Atomic Power Station in Ukraine.

There is a formal agreement for a partnership for quality between labor and management in this office, and a strong commitment to the concept by leadership of our employees' union and some NE managers and employees. In some respects, NE is a model for the quality organization that is needed to manage nuclear technology, and staff members are often called upon to provide advice to other organizations or to the Secretary on important nuclear matters.

NEED: CORPORATE INSTEAD OF BUREAUCRATIC MANAGEMENT - THE UNITED STATES NUCLEAR MATERIAL RESOURCES CORPORATION

The Office of Nuclear Energy is one of twenty organizations in DOE, of which eight have responsibility for nuclear technology. None of the other seven has a strong partnership between labor and management and there is little overall coordination. All are dominated by political leadership which changes every few years thus there is no assurance of the continuity that is essential to sustained quality.

Success of some NE programs depends on other DOE offices; for example, supply of RTGs that provide electricity for instruments aboard vehicles that travel into deep space, such as Galileo at Planet Jupiter, requires Plutonium-238 which was produced in the past by DOE's Office of Defense Programs. But Defense Programs discontinued production of this material in 1983. NE programs are no worse off than those of the U.S. Department of Defense which will need tritium to replace that in our strategic nuclear deterrent lost through decay - but there is no capability for production of this material and no firm programs are underway to establish the capability, despite major initiatives by former Energy Secretaries Don Hodel and James Watkins. NE is also no worse off than utilities operating nuclear power plants and relying on DOE for acceptance of spent nuclear fuel, and hospitals and other commercial organizations that need sites for stabilization and isolation of low-level radioactive wastes. DOE has sites that would be ideal for management of spent nuclear fuel, for stabilization and isolation of low level radioactive wastes, and for production of nuclear materials that are needed for important national programs. But decisions must be made to do the right thing.

The other keynote speaker at the conference last April in Moscow was Viktor Mikhailov, Minister of Atomic Energy of Russia, who talked about problems facing nuclear power plant operators that had supplied electricity free to customers under the centrally controlled economy of the Soviet Union. Under the present market economy in Russia, customers are expected to pay, but they are not. He said that workers usually go for months without pay, and that needed reactor safety upgrades without any money is very difficult.

I was interested in Dr. Mikhailov's remarks - but most interested in the fact that only he and I, of all speakers at the conference, represented nations where nuclear technology is managed by a cabinet-level government department.

The Department of Energy is a government bureaucracy with responsibility for managing very complex nuclear technology. Fifty-four years ago, at the dawn of the nuclear age, leaders of the United States Government wisely determined that a government bureaucracy could not carry out the tasks needed for this complex technology; that corporate experience, know-how, and organizational strength would be needed.

The DuPont Company, a U.S. corporation with an almost 200-year history of excellent safety experience using complex technology involving hazardous materials, was selected to design, construct, and operate nuclear reactors and reprocessing plants for production of plutonium at the Hanford Engineer Works, in Washington State - and carried out that mission with great success. Management practices of this corporation emphasized best possible input from knowledgeable employees and other knowledgeable sources prior to corporate decisions for new ventures - and full input from employees on any safety concern, including regular meetings among employee groups for full resolution of these concerns.

During the late 1940s the Soviet Union designed, constructed and operated facilities for similar effort. It's interesting to compare experiences at the U.S. corporate and Soviet bureaucratic facilities. At the U.S. plants, maximum radiation exposure to personnel was about one REM per year. At the Soviet plants, average radiation exposure per person was 150 REM per year; maximum exposure was more than 300 REM per year. (Five REM per year is the maximum allowable radiation dose to U.S. workers; 500 REM is a lethal dose to most people.)

And less we attribute all of the difference to Soviet Communism, note that average radiation exposure to personnel at the Nuclear Fuels Services, Inc., reprocessing plant in the U.S. during its last year of operation was almost eight REM per year, maximum exposure was sixteen REM, and exposures were rising exponentially. The NFS reprocessing plant design was the product of bureaucratic folly of the U.S. Atomic Energy Commission. Also, the Atomic Energy Commission issued a construction permit for a commercial nuclear facility in which a credible accident could have released gross quantities of highly radioactive materials - and some senior executives of the Department of Energy planned a demonstration incorporating that facility, disregarding concerns of experienced employees. Fortunately, President Ronald Reagan said NO!

Because of the outstanding success of its operations at the Hanford Engineer Works, President Harry Truman requested that The DuPont Company design, build, and operate the Savannah River Plant (SRP) for the United States Atomic Energy Commission (AEC) to produce nuclear materials for defense and peaceful applications - and The Company did so, again with great success.

Twenty-five years ago SRP was being operated by DuPont to produce nuclear materials under direction, leadership and oversight of the AEC Savannah River Operations Office (SROO). Virtually all facilities were in operation and most of the materials produced were for peaceful applications including plutonium-238 for exploration of deep space; tritium was produced to maintain the strategic nuclear deterrent. Near world-best safety records (best ever DOE and AEC safety records) were being established; reactor operation was under continuing review of the Advisory Committee for Reactor Safeguards and judged as safe; protection of workers and off-site personnel from radiation, and that part of the environment important to life was outstanding. Operation had just started in a new radiochemical processing facility for recovery of gram quantities of Californium and milligram quantities of Berkelium and Fermium with highly advanced pressurized chromatographic cation exchange technology. Plutonium scrap from SRP operations was recovered immediately and there was no excess accumulation; scrap from other AEC sites was being delivered to SRP for recovery there. Highly radioactive products of fission processes, so-called high level radioactive wastes, were being stored safely in a manner that would permit later recovery of the valuable materials for beneficial use. Studies were underway for advanced concepts for commercial nuclear fuel reprocessing based on world-best experience that would have provided increased safety and environmental protection and greatly increased safeguards, material controls, and proliferation resistance. Full potential of SRP land was achieved by the AEC and the U.S. Forestry Service through the planting, site-wide, of pine trees and harvesting by commercial firms for paper and wood products. (Prior to acquisition by AEC, most of the land was devoted to very marginal cotton farming and hunting.) There were extensive and unique programs for environmental and ecological research. Total employment at the plant and laboratory, including DuPont corporate management in Wilmington Delaware, was about 10,000.

Five years ago at SRP there were almost 30,000 employees; the entire complex was devoted to military activities under DOE Office of Defense Programs management and leadership, and Defense Nuclear Facility Safety Board oversight; virtually nothing was in operation. Another interesting comparison of corporate versus bureaucratic management and oversight.

The studies that would have provided an excellent model for worldwide management of spent nuclear fuels had been completed years earlier but were rejected by bureaucratic managers of the Energy Research and Development Administration and never made available for consideration in policy decision processes of The Congress or The President.

The United States created world-best nuclear technology for its weapons programs through U.S. corporate know-how, expertise, and organizational strengths, with input from research and development from government, university, and commercial laboratories.

Some of this technology was adapted for commercial nuclear programs; nuclear power plants were deployed by U.S. corporations with great success throughout most of the world.

Uranium enrichment services were made available by the United States, but under government bureaucratic management which resulted in questionable marketing policies, wasted expenditures of several billions of dollars, and limited advancements in enrichment technology. Transfer of uranium enrichment programs from bureaucratic to corporate management should lead to improvements, particularly if the U.S. Enrichment Corporation incorporates advanced quality principles in its management practices.

Responsibility for disposition of commercial spent nuclear fuel and spent fuel of U.S. origin from abroad was initially assigned to an AEC industrial contractor with world-best experience in spent fuel management - but misinformation from the AEC led to commercial reprocessing based on technology that had failed in AEC operations.

Upon full recognition of failure of commercial reprocessing, and under the leadership of AEC Chairman Dixy Lee Ray, experienced senior managers and competent technical staff undertook study with input from U.S. nuclear utilities and nuclear companies, and reassigned efforts for development of technology for spent nuclear fuel reprocessing and recycle to build on successful experience. But competent managers and technical staff were discarded as AEC responsibilities were transferred to the Energy Research and Development Administration, and reprocessing development leadership was transferred back to the source of flawed technology.

Continuing bureaucratic folly since that time has led to indefinite accumulation of spent fuel at commercial nuclear power plants and DOE facilities, indefinite accumulations of separated plutonium, waste of many billions of dollars, plans to create geologic deposits of accessible weapon usable materials that could pose dire threats to future populations, loss of credibility of nuclear technology in the United States, and U.S. nuclear policies that constrain collaborative research and development among nations that is needed to assure best controls and safeguards for nuclear materials and technology.

The twenty-two-year moratorium on advancement of nuclear technology by a former world leader in the technology has constrained development of the systems that would have provided best assurances of national and international safeguards - and best assurances for national and global security. We live on a more dangerous planet because of the vacuous nuclear policies of the United States.

The indefinite accumulation of plutonium-containing spent fuel, or its "disposal" in geologic formations is a proliferation threat that increases with time, as short lived fission products decay. The short-lived fission products in fresh spent fuel provide intensely penetrating radiation that limits access to plutonium, protects it from diversion, and makes reprocessing difficult. Some of the short-lived fission products are gaseous, and their release during reprocessing provides an indication of reprocessing activity - overt or covert. After decay of these short-lived radionuclides, separation of plutonium from spent fuel could be carried out in readily available processing equipment - such as that used for processing of dairy products - and could be carried out in a barn, garage or basement, without significant danger to those carrying out the operation and without significant risk of detection. Safeguards in-perpetuity could limit this threat, but is not feasible and is not provided for by provisions of the U.S. Nuclear Waste Policy Act.

In order to avoid this perpetual threat, plutonium and other potentially weapon usable materials must be removed from spent fuel, prior to its indefinite accumulation or "disposal" in a repository. The radiochemical processing steps for recovery of potentially weapon usable materials, and other materials that have potential for beneficial use - usually called reprocessing - can be carried out in well designed facilities in secure locations, without a significant proliferation threat. All nations that use light water reactors recognize this need, except the United States and Sweden, and Sweden plans to discontinue use of nuclear power. Sweden requires a period of 100,000 years for "engineered containment" of the spent fuel; the United States requires only 10,000 years. In 100,000 years, plutonium remaining in the planned repository at Yucca Mountain in the United States would, after decay, be sufficient for about 7,000 nuclear weapons. There would be no protection of this material by intensely radioactive fission products, thus the plutonium would be accessible through simple mining and chemical processing techniques.

In best designs for reprocessing and spent fuel recycle, such as the "Spent LWR Fuel Recycle Complex" design integration study completed by DuPont in November 1978, plutonium would be "in-process" and contained within heavily reinforced concrete structures until removed from the well-safeguarded complex in the form of mixed-oxide fuel elements or assemblies. There would be no accumulation of separated plutonium, and diversion of plutonium from an in-process stream for inappropriate use would not be feasible, or even possible without significant modification which would be easily detected by electronic surveillance equipment or inspection. The final product fuel assemblies would be shipped under appropriate safeguards for recycle into reactors, where neutron irradiation would restore short-lived fission products that would preclude any credible threat of a plutonium diversion. The ultimate threat of proliferation from commercial nuclear power plant operation is greater from indefinite accumulation or unsafeguarded geologic "disposal" of unreprocessed spent fuel than from well managed, well engineered, well safeguarded reprocessing in a secure location.

The misunderstanding within the United States about the proliferation threat of reprocessing of spent fuel from commercial nuclear power plants developed as an aftermath of India's detonation of a nuclear explosive manufactured from plutonium separated in a reprocessing plant built from reprocessing technology unwisely supplied by a national laboratory and an engineering firm in the U.S. - and produced in a so-called "research" reactor unwisely supplied by Canada. India's Cirus Reactor is similar to one operated by Canada to produce plutonium for the U.S. nuclear weapon program under a U.S.-Canada mutual defense agreement.

There has never been a diversion or known attempt for diversion of plutonium from reprocessing of spent fuel from reactors dedicated to production of electricity, and it is unlikely that there will be, simply because it is easier to produce weapon usable materials in facilities dedicated to that mission. Pakistan and South Africa demonstrated the ease of undetected weapon material production through enrichment of uranium; India produced plutonium for its nuclear weapon program because India had been provided with a weapon grade plutonium production complex, and was allowed to operate that complex without international controls on plutonium.

Deferral of reprocessing by the United States has not resulted in other nations following that example; failure of the United States to proceed with reprocessing of commercial spent fuel and use of plutonium with its "world-best" technology, and failure to share that technology with other nations, has decreased global security, limited nuclear material controls and best safeguards, and led to serious and difficult problems in the United States.

It is time for leaders of the United States to reevaluate systems for management of nuclear technology in the U.S. and abroad, discard those that have led to failures, and hold on to, reinstate, or install those that have worked best. As an initial idea, the United States should follow the lead of most other nations and create a corporate structure "The United States Nuclear Material Resources Corporation" for management of government nuclear programs that would have the continuity that is essential for quality, and build the corporate memory and organizational strengths that are essential for appropriate management of complex technology.

This corporation would:

(1) take responsibility for three major DOE nuclear sites, three major laboratories and other sites and facilities as appropriate, and continue programs at these sites for environmental and ecological research and other productive uses;

(2) with government, corporate, or other funding as appropriate, operate nuclear reactors and other facilities at these sites for production, fabrication, and recycle of nuclear materials needed for defense, space exploration, medicine, industry, research, energy development, and other purposes, or assure that these materials are available from competent private corporations;

(3) through transfer of the DOE Nuclear Waste Fund, continuing payments into this fund, and funding as now planned for environmental management, manage the by-products of nuclear fission processes (so-called high level radioactive wastes) at these sites, receive and store spent nuclear fuel, recover actinide elements and fission fragments and make them available for beneficial uses, accept low level radioactive wastes for isolation and stabilization at appropriate charges from medical, industrial, educational, government, and other organizations, and continue cost beneficial environmental management efforts that are needed to ensure health and safety for all life forms;

(4) assure that all potentially weapon usable nuclear materials (uranium, plutonium, neptunium, etc.) in excess of military requirements are used in nuclear reactors to recover energy values and render these materials inaccessible for diversion for unintended uses;

(5) in collaboration with organizations in the U.S. and abroad, including shared funding and shared potential for profit, carry out and/or sponsor research, development, and demonstration of advanced nuclear reactor and fuel cycle concepts;

(6) in conjunction with the National Academy of Sciences, National Academy of Engineering, and Institute of Medicine, and others as appropriate, function as a source of accurate information on nuclear technology for American Citizens, political leaders, other government organizations, and commercial organizations; and

(7) in partnership with the U.S. Nuclear Regulatory Commission, facilitate the development of improved technical bases that will service as the basis for cost effective regulatory standards that will assure best safety for nuclear activities.

The ten - fifteen members of the Board of Directors of this corporation should be selected by The President, with staggered terms. Half of the members should be career employees of the corporation; others should be selected from industry, universities, the military, political leadership, unions, laboratories, or space programs. This corporation should be managed with best and most advanced quality principles, including full partnerships between labor and management.

NEED: BETTER SYSTEMS FOR DEVELOPMENT OF POLICIES -
THE UNITED STATES NUCLEAR TECHNOLOGY POLICY BOARD

During the summer of 1995, an American Nuclear Society (ANS) Special Panel on Plutonium issued a report which provides an excellent framework for actions that would lead to more viable use of nuclear technology in the United States - but the report is being ignored.

Since 1974, nuclear policies have been developed in the United States through study by the U.S. Congress Office of Technology Assessment (OTA), by White House staff , and by the Department of Energy - often with little input based on experience. White House and DOE developed policies change, often drastically, from one administration to the next or as a result of changes in White House or Department of Energy senior staff, thus there is little chance for the continuity that is essential for quality. Quality of OTA staff work was such that the organization has been abolished. The result of U.S. nuclear policies over the past 22 years is a nuclear vacuum, which is dangerous and expensive, particularly for a nation that had been and should be a world leader in nuclear technology.

Members of the ANS special panel for plutonium were appointed by the President of ANS. If they had been appointed by President Bill Clinton with the advice and consent of Congress - same panel, same members, same recommendations - the U.S. would be moving forward toward safer, more cost effective, and more beneficial use of nuclear technology.

The Congress should enact and The President should sign legislation creating "The United States Nuclear Technology Policy Board" whose seven members would be appointed by The President with the advice and consent of The Congress. The members would be individuals with extensive experience in or highly knowledgeable of nuclear technology such as Nobel Laureate Glenn Seaborg or Harvard Physics Professor Richard Wilson; a senior technical executive of an industrial organization with a strong commitment to safe operations and quality, such as DuPont; a political leader such as Senator Pete Domenici or Senator Frank Murkowski; NTEU National President Bob Tobias or AFL-CIO President John Sweeney; Nuclear Energy Institute President Joe Colvin; American Medical Association Vice President Jim Allen; former American Nuclear Society President Alan Waltar; former General Electric Nuclear Division vice-president Bert Wolfe; designees of the Secretary of Defense and the National Aeronautics and Space Administration; former Ambassador Tom Pickering; and CIA Director John Deutch. This board would meet periodically to review government and commercial nuclear programs and lay out long term nuclear strategies for the nation.

NEED: MANAGEMENT FOR BENEFICIAL USE OF ALL
NUCLEAR MATERIALS INSTEAD OF THEIR DISPOSAL

Funding for the research that led to the discovery of nuclear fission was provided by a chemical company, not for military or energy uses but for radiochemicals for medical applications. In the U.S. today, medical use is the greatest beneficiary of nuclear technology.

The narrow focus on a single product of most uses of nuclear fission and categorization of all other potentially valuable products as "waste" has led not only to denial of great benefits, but also to the accumulation of potentially dangerous "wastes" - whose "disposal" in geologic formations is neither politically acceptable nor technically responsible. Unreprocessed spent fuel from commercial nuclear power plants contains potentially weapon usable materials that would be available and accessible for recovery and use or misuse for millions of years. Recycle and efficient use of nuclear materials for energy production would not only reduce or eliminate this threat by maintaining potentially weapon usable material in inaccessible form, but would supply with minimum adverse impact on the environment much needed electricity for hundreds of thousands of years, and valuable by-products for many beneficial uses.

In the first use of nuclear fission, during World War II, only plutonium was recovered; unused uranium and other actinide elements and fission products were discarded as waste. Later, because of shortages, uranium that had been discarded was recovered and recycled for production of plutonium. Beginning in the late 1950s, neptunium-237 was recovered from AEC radiochemical processing plant waste streams and irradiated to produce plutonium-238 for a possible weapon application. The weapon application did not turn out to be feasible and large quantities of plutonium-238 were available - and used to supply power for instruments aboard space craft. Now plutonium-238 is essential to exploration of deep space. Neptunium-237 has a high fission cross section and thus has potential for use as a nuclear explosive - and with a two-million year half-life, this material would be available for potential misuse for many millions of years unless recovered and irradiated for production of plutonium-238 or electricity.

Cesium-137, the major intensely radioactive constituent of spent fuel and high level radioactive waste, was recovered from waste streams at Hanford in order to reduce heat generation in tanks intended for long-term isolation of wastes, and now has great value as an intensely radioactive source. The need for irradiation of food to prevent E-Coli infections, and of blood plasma and blood plasma products to prevent spread of AIDS type viruses, could lead to greatly increased demands for cesium-137.

All products of nuclear fission processes have potential value; many have great value. By recovering and using these products for beneficial purposes, adverse impact on the environment is diminished, and potential for inappropriate diversion and misuse for destructive purposes is reduced or eliminated.

NEED: OPENNESS, AND ACCURATE INFORMATION TO THE PUBLIC

The openness that Secretary Hazel O'Leary brought to the Department of Energy is a great need for nuclear technology, similar to that brought to the U.S. Atomic Energy Commission by Chairmen Glenn Seaborg and Dixy Lee Ray. We must institutionalize that openness.

One of the most important benefits of our democratic government is the opportunity it affords for full participation in major decision processes by all of its citizens. But this benefit can only be realized through an informed citizenry. It is essential that systems and processes be installed that insure best possible, timely and fully understandable information is provided to the public on all aspects of nuclear technology.

NEED: STRONG COLLABORATIVE RESEARCH AND DEVELOPMENT PROGRAMS AND SHARED TECHNOLOGY WITH OTHER NATIONS FOR BEST SYSTEMS FOR BENEFICIAL, EFFICIENT AND SAFE USE OF NUCLEAR TECHNOLOGY

Nuclear technology and nuclear materials, and information on the great benefits of their use, are available to virtually every nation on this planet. Growing populations, developing prosperity among nations, and increasing awareness of global interactions of energy with the environment will lead to increased demands by leaders and citizens of all nations for full and efficient use of nuclear technology. A safe and secure planet with a healthy environment will require best assurances of safeguards against diversion of potentially weapon usable materials, and safe operations to preclude release of radioactive materials into global atmospheres. These assurances can only be achieved through shared technology and collaborative research, development, and demonstration among nations. Denial of nuclear technology to nations for peaceful uses of nuclear technology will ultimately limit best safeguards, best systems for environmental protection, and most efficient and cost-effective use of nuclear technology.

CONCLUSIONS

The Systems that are needed by the United States to insure that safe, non-polluting and cost-effective nuclear technology is fully utilized for the benefit of its citizens and all mankind include:

(1) Adoption of partnership processes between labor and management, and between regulators and regulated organizations for assurance of safety and resolution of differences;

(2) the United States Nuclear Material Resources Corporation, a corporate structure with full commitment to best quality principles, including continuity, for management of government nuclear programs;

(3) the United States Technology Policy Board, whose members are appointed by The President with the advice and consent of The Senate that would meet periodically to recommend or establish long-range nuclear policies;

(4) management for beneficial use of all nuclear materials instead of their disposal;

(5) openness and full and accurate information to the public on all activities; and

(6) strong programs for research, development and demonstration in collaboration with other nations for advanced nuclear systems for beneficial, efficient, cost effective and safe use of nuclear technology, including improved safeguards and other assurances to preclude diversion of nuclear material for inappropriate uses.

The nuclear fission process provides tens of millions of times the energy output of combustion of conventional fuels such as coal, natural gas, petroleum, wood, and biomass and can meet energy needs with much less adverse impact on the environment than any other energy source. The valuable products of the fission process have potential for great benefit, but good management of these products is needed, not plans for their disposal.

With nuclear technology, much more can be done with less. But best systems are needed for cost-effective management; assurances of safety, safeguards and environmental protection; and development of policies for its use. We must assure that best systems are provided, maintained, and improved, and that there is full collaboration among nations for beneficial use and not misuse of nuclear technology.

GLOBAL NUCLEAR ENERGY/MATERIALS MODELING IN SUPPORT OF LOS ALAMOS NUCLEAR VISION PROJECT: LONG-TERM TRADEOFFS BETWEEN NUCLEAR- AND FOSSIL-FUEL BURNING

R. A. Krakowski

Systems Engineering and Integration Group
Technology and Safety Division
Los Alamos National Laboratory
Los Alamos, New Mexico 87545

ABSTRACT

A global economics/energy/environmental (E^3) model has been adapted with a nuclear energy/materials model to understand better "top-level", long-term trade offs between civilian nuclear power, nuclear-weapons proliferation, fossil-fuel burning, and global economic welfare. Using a "business-as-usual" (BAU) point-of-departure case, economic, resource, and proliferation-risk implications of plutonium recycle in LWRs, as well as greenhouse-gas-mitigating carbon taxes and a range of nuclear-energy costs (capital and fuel), have been examined. After describing the essential elements of the analysis approach being developed to support the Los Alamos Nuclear Vision Project, preliminary examples of parametric variations about the BAU base-case scenario are presented. The results described herein represent a sampling from a collection of more extensive results. The primary motivation here is: a) to compare the BAU base case with results from other studies; b) to model on a regionally resolved global basis long-term (to year ~2100) evolution of plutonium accumulation in a variety of forms under a limited range of fuel-cycle scenarios; and c) to illustrate a preliminary connectivity between risks associated with nuclear energy and fossil-fuel burning (e.g., the relationship between nuclear proliferation and greenhouse-gas accumulations).

INTRODUCTION

The "tension"[1] between nuclear-weapons (NWs) and nuclear-energy (NE) uses of plutonium has generated a deeply divided and evolving debate over the growing commercial

Technology for Global Economic and Environmental Survival and Prosperity
Edited by Kursunoglu et al., Plenum Press, New York, 1997

201

(spent fuel) and military (retired NWs) inventories of this material[2]. The recent releasing of weapons plutonium has highlighted a plutonium legacy and the associated "clear and present danger"[3]. The broader debate over the best way of dealing with the plutonium issue, however has centered primarily on reprocessing of the larger inventories of plutonium in spent nuclear fuel. Over the years, the rationale for commercial reprocessing and plutonium recycle has moved[4]: a) from primarily economic and energy-security arenas (up to the mid-1970s); b) to less-strategic justifications based on improved management of NEs radioactive wastes (up to the mid-1980s); c) to the present stance centered on interim-storage *versus* direct-disposal options that are based primarily on long-term environmental and (once again) energy-security considerations. Arguments against closing the nuclear fuel cycle through reprocessing are based largely on fears of accelerated increases in proliferation potential as inventories of civilian plutonium grow (*e.g.*, each tonne of plutonium in any form is equated to ~100-200 NWs), with the evolution of this situation being driven by processes that are claimed not to be economic for decades to come. Contemporary arguments supportive of plutonium recycle have focused[5] primarily on non-economic issues that generally invoke rationale that are: a) environmental (*i.e.*, reduced bio-toxicity of disposed wastes); b) resource-based (*i.e.*, uranium conservation through recycle of both uranium and plutonium in LWRs); c) strategic (*i.e.*, energy independence and options flexibility, particularly for nations without large resource endowments); d) political (*i.e.*, proliferation risks are claimed to be reduced by reprocessing scenarios that minimize accessible inventories of separated plutonium); and e) risk-minimizing (*i.e.*, technology footing of reprocessing is firmer than direct disposal).

Among the elements contributing to the complexity of the reprocessing/recycle debate are: a) the long-term (*i.e.*, "plutonium mine") *versus* short-term (*i.e.*, spent-fuel *versus* separated/stockpiled plutonium forms) characterization of the proliferation risk; b) regionalization of growths in population and associated energy demand; c) globalization of energy supply and environmental impacts; and d) the relationships between security of energy supply, economic well being, and regional propensities for nuclear proliferation. A central issue in this debate is the degree to which mankind should and can restrict the exploitation of a major sustainable energy source in the pursuit of a world with an acceptably low NW proliferation risk and optimum equity in energy utilization and human welfare (*e.g.*, prosperity with security).

The Nuclear Vision Project[6] at Los Alamos is exploring alternative global (nuclear) energy futures over the next $\geq$ 50 years. These futures are being probed within the contextual realities of: a) significantly increased population and energy demand in east Asia; b) the implications this increased population coupled to increased *per-capita* energy consumption has on global energy security; c) the role NE potentially can play in this projected growth, particularly in relieving resource and environmental strain related to the use of nonrenewable fossil fuels; and d) the implications of this growth on the nuclear-energy exploitation *versus* nuclear-weapons proliferation (*e.g.*, "MW *versus* Mtonne") issue elaborated in Ref. 2. The charting of possible roadmaps leading to desirable future endstates requires at least five essential elements that are incorporated into the Los Alamos study: a) nuclear weapons; b) nonproliferation; c) nuclear materials (NM, *i.e.*, fissionable isotopes of uranium and plutonium, more conventionally referred to as "special nuclear materials"; NM in the context used here does not include natural or low-enrichment uranium) inventories; d) energy and environment as driven by economic considerations (E^3); and e) institutional and public acceptance of large technical systems that portend large risk and require unusually high levels of reliability[7] in related operations. Additionally, an overarching determinant of future global endstates is the relationship between global security and energy security, and the way in which that relationship is impacted by underlying connectivities between nuclear weapons, nuclear materials, and nuclear energy. Important linkages exist between the NW/NM/NE elements of the Nuclear Vision Project and human activities and consequences related to the pursuit and use of other (both fossil and nonfossil) energy sources that are essential in the search for increased equity in human welfare. In approaching these complex and interconnected issues, the Los Alamos Nuclear Vision Project is comprised of three main components: Workshops and Outreach Activities that are supported by Analyses; this report summarizes the (evolving) Analysis component.

APPROACH

The Institute of Energy Analysis (Oak Ridge Associated Universities, IEA/ORAU) Long-term Global Energy Economics Model[8], because of its scope, transparency, and historically broad use[9-13], was adopted as a computational platform for the Nuclear Vision Project. Other E[3] models, however, have and are being considered[14-19]. An operational description of the ERB (Edmonds, Reilly, Barns)[8] model, as it is being modified and applied to the Los Alamos study, is given in Ref. 20. The nuclear economics and fuel-cycle model being evaluated "under" the ERB model is also elaborated in Ref. 20, with Ref. 21 describing the recent evolution of the proliferation-risk model being evaluate in conjunction with global NM flows and NE economics (and related regional and global market shares). The NE model being developed for use with the ERB model is based primarily on NM flows within the civilian nuclear fuel cycle. A more detailed description of the analytic approach and related interim results are given in Ref. 22, from which most of the material reported herein derives.

Global E[3] Modeling

Overview

General Considerations/Background. The desire and need to look forward into the world energy future combines with an inability to predict the future to prescribe a narrow intellectual tightrope along which any modeler of long-term energy futures must tread.[10] Figure 1 depicts a model as a generalized structure on which the interactions between a range of human activities are examined. Assumptions about relevant human activities provide input to the model, and output is generated in the form of an array of possible impacts (e.g., in the present context, GHG emissions/accumulations, NM inventories, related proliferation risks, economic/energy intensity and development, etc.). The fact that alternative human activities can lead to large consequential differences contributes to the large uncertainties in "forecasting" future courses of events. As is indicated on Fig. 1, an understanding or appreciation of these uncertainties is developed by the modeling process through various levels of analyses that includes error [both of omission and (model) distortion or imperfection] analyses. Once consequences (e.g., of GHG accumulations, form and magnitudes of NM inventories, economic impacts of use or restrictions of specific technologies, resource depletion and/or inequities and related security implications, etc.) emerge from the global E[3] analyses, recommendations are formulated and placed in a que of many impacts vying to inform decision makers aid in and the creation of regional and global policy. In the context of the Los Alamos Nuclear Vision Project, the modified ERB model produces certain results, guidelines, and/or estimates that, along with many other inputs, are folded back into the investigation underway. Large uncertainties characterize these results, and the explication and understanding of the large uncertainties that link possible causes and effects identified by the process described in Fig. 1 are often poorly resolved. In spite of these uncertainties, scenario building using structured models offer a number of attractions that include[10] an ability:

- to test sensitivities of a given "forecast" to input assumptions;
- to explore behaviors under extreme or unlikely conditions;
- to assess relative importance of alternative assumptions;
- to represent explicitly specific tradeoffs [*e.g.*, GHG *versus* proliferation risk *versus* technology and economic risks associated with the pursuit of advanced (*i.e.*, renewable, sustainable) energy technologies];
- to understand or explicate uncertain or unresolved linkages (*e.g.*, human activity → GHG emissions; GHG emissions → GHG concentrations; GHG concentrations → temperature change; temperature change → geophysical impacts; geophysical impacts → effects on the welfare of humans and others; similar chains or linkages can be generated for the accumulation of nuclear materials, related proliferation risks, and the time scales required for human commitment).

Clearly, explicit investigations that are required to establish important linkages listed in the last item are not within the scope of the Nuclear Vision Project, and the extensive work that is ongoing in understanding global climate change will be used, as is appropriate and necessary.

General Approaches. A wide range of approaches to global energy modeling to provide behavioral forecasting tools have been reported over the last 2-3 decades. Three generic modeling approaches can be identified[10]. First, single-pass analyses establish intuitive relationships between drivers that influence global energy production and use; mismatches between energy production and use in these generally analytic models are resolved by judgment or the assumption that "backstop" energy technologies[23] (*e.g.*, very large resources become available only at high prices) are available to fill the gap. A second generic approach is embodied in the computer-based Linear Programming (LP) models[14,17,23,24] that, upon entry of large amounts of input information, generates either optimized (*e.g.*, GNP) or constrained (*e.g.*, GHG-limited) forecasts. A hybrid between Linear Programming, input-output, and energy-accounting models represents a third class of global E[3] models[14,19]; this approach is generally complex and not easily used. The ERB model[8] adopted for the Los Alamos Nuclear Vision Project is base on a behavioral market equilibrium that internally balances energy production and usage; while simplified in comparison to the LP-based or hybrid models, the ERB model was judged to be targeted adequately to the (early) needs of the Nuclear Vision Project, is available for use by the public, is adaptable to modification, and is technically transparent and well documented.[8,10]

ERB Model

The ERB model was developed nearly 25 years ago at the IEA/ORAU under contract to the DOE for the CO_2 Research Division and has been adapted to examine CO_2 emissions by several institutions, including EPA, MIT, EPRI, and GRI. Although an earlier version is available electronically through IEA/ORAU, PNL[25] supports more recent versions. The

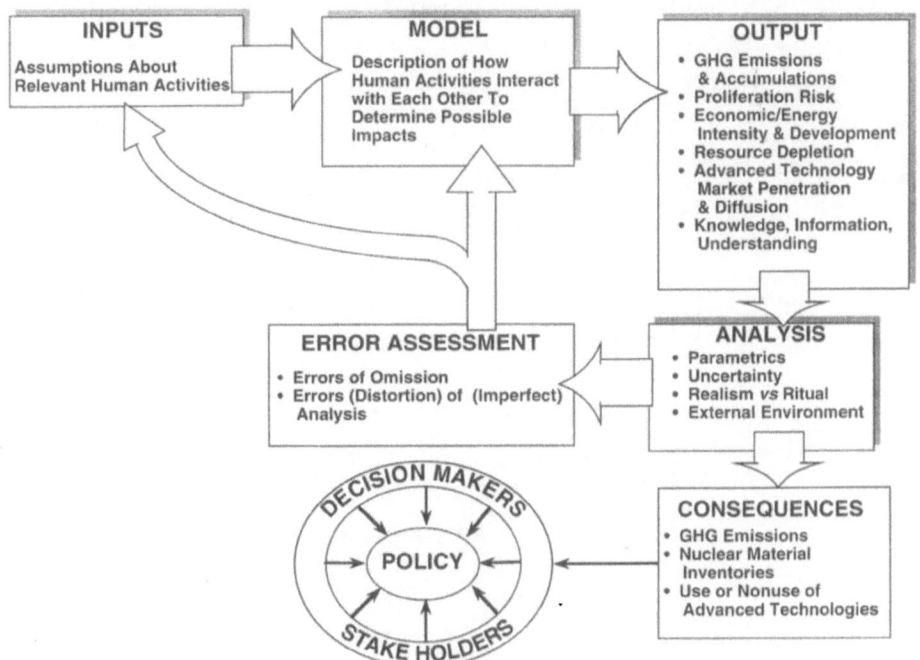

Figure 1. Fundamental components and functionalities of a generic E^3 (energy/economics/environment) model; the examples given under "Output" and "Consequences" pertain to the specific study reported herein.

ERB model is comprised of four main parts: supply, demand, energy balance, and GHG emissions (a postprocessor). Supply and demand are determined for six primary energy categories: oil (conventional and nonconventional); gas (conventional and nonconventional); solids (coal and biomass); resource-constrained renewables (hydroelectric and geothermal); nuclear (fission, with fusion being included as a form of solar energy[26]); and solar (excluding biomass, includes solar electric, wind, tidal, ocean thermal, fusion, and advanced renewables; solar thermal is included under "conservation"). The energy-balance module assures that supply equals demand in each of nine global regions, with primarily electrical energy assumed not to be traded (*e.g.*, assumed to be generated and used within a given global region). The nine global regions are modeled over a ~100-yr time horizon, with the NE status of the regions being tracked by the NE/NM model that has been added to the ERB model.[20] While the GHG emissions are computed after global economic equilibrium and energy balance is achieved for each to the nine 15-year time steps, the nuclear component, as modified for use by the Nuclear Vision Project[20,22], must be evaluated integrally with the iterative approach to economic and energy equilibrium that forms the heart of the ERB model.

A qualitative description of each of the main ERB modules [demand, supply, energy balance, GHG emissions, and nuclear (costing, nuclear material flows, proliferation risk] is given in Ref. 22, Ref. 8 gives a detailed description, and Refs. 20 and 21 give an operational description of the modified model, as used in the Nuclear Vision Project application. Figure 2 is a simplified graphical description of the computational algorithm used the ERB model and includes the modifications made to support the Nuclear Vision Project; this diagram is an extension of that given in Ref. 8.

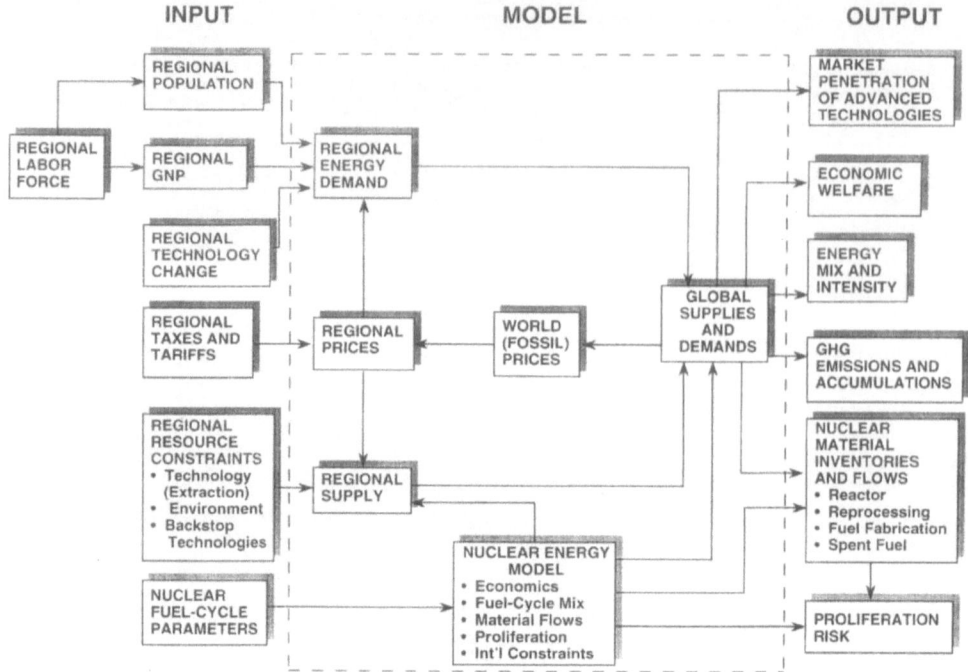

Figure 2. Structure of ERB global E[3] model[8] as adapted for use in the Los Alamos Nuclear Vision Project; four components comprise the ERB model: energy demand; energy supply; energy balance; and GHG emissions (a postprocessor action); the relationships between inputs and outputs, as well as addition of higher-fidelity nuclear model[20] and proliferation-risk model,[21] are shown.

Nuclear Energy Model

The nuclear model developed and implemented for operation "under" the ERB global E[3] model performs three main functions: a) determines a "top-level" cost estimate in terms of a cost of electricity (COE, mill/kWeh or $/GJ) that is reformed into the Leontief coefficients[27] used to determine costs and market shares, as described above; b) tracks the flow of key elements throughout the nuclear fuel cycle [*e.g.*, natural uranium, low-enriched uranium, plutonium, and spent fuel] for use in subsequent waste and proliferation-risk assessments; and c) performs a multi-attribute utility (MAU) analysis to estimate proliferation risk associated with the civilian fuel cycle. The costing and material-stream flows are described in Ref. 20, whereas the MAU-based proliferation-risk assessment methodology is elaborated in Ref. 21.

Before costs, NM flows/inventories, and proliferation risk can be estimated, however, characteristics of the fuel cycle must be specified. The nuclear model reported in Ref. 20 is based only on the U/Pu cycle, as implemented in each global region at each time interval by an economically determined ratio of electrical generation provided by thermal-spectrum Light-Water Reactor (LWR) and fast-spectrum Liquid-Metal Reactor (LMR) systems. The LWR in a given global region operates along an exogenously determined MOX recycle trajectory that exponentially transcends from an initial MOX core fraction to a final MOX core fraction with a specified time constant. The LMR system, if economics and technology diffusion time constraints allow, is introduced with a preassigned breeding ratio. Efforts to manage plutonium inventories through regional and temporal control of MOX recycle fraction and (LMR) breeding ratio to date have not been extensive, within the context of the present modeling effort. Generally, plutonium is assumed to flow freely between global regions, where deficits in some regions are corrected by flows from regions with excess plutonium; the global plutonium inventories, however, are constrained to remain positive. Detailed plutonium balance and control remains for future work and higher-fidelity nuclear and costing models. Specifically, implementation of interregional NM flow constraints, breeding ratios driven by NM inventory and need requirements, and cost- and/or sanction-based selections of MOX recycle fractions represent important areas of future work.

Figure 3 illustrates a generic fuel cycle that has been constructed from a series of building blocks identified with key fuel-cycle processes. The simplified species-resolved mass balances described in Ref. 28 based on the kind of input-output analysis depicted on the bottom of Fig. 3 are used to model material flows. Unit capital and operating costs are applied to each of the processes depicted in Fig. 3, from which a fuel-cycle cost for the entire system is estimated. Plutonium flows and accumulations are monitored for each region as a function of time, with reactor plutonium, separated plutonium in reprocessing and fuel fabrication, and accumulated in spent fuel the four major categories being tracked. These plutonium inventories are used as part of the time-dependent proliferation-risk assessment associated with each global region, for a specified set of exogenously determined nuclear and ERB parameters.

Proliferation-Risk Model

Proliferation Issue

While the present study follows past practice[29-35] and focuses quantitative attention almost completely on the NM streams within the civilian nuclear fuel cycle, the broader perspective outlined in Refs. 29 and 36 has been charted on Fig. 4. The primary purpose of Fig. 4 is to suggest a methodological framework with which to assess proliferation risk that pushes beyond earlier approaches centered primarily on NM availability (*e.g.*, total inventories) and related proliferation risk. As discussed in Refs. 29 and 36, the decision on the part of a given entity to proliferate and pursue a path towards NW "breakout" can be aggregated into five top-level inputs or drivers:

- INCENTIVES to build, possess, integrate, maintain, and threaten to use nuclear weapons;
- DISINCENTIVES to build, possess, integrate, maintain, and threaten to use nuclear weapons;
- TECHNICAL MEANS needed to build, possess, integrate, maintain, and threaten to use nuclear weapons;
- MATERIAL AVAILABILITY/ACCESS needed to build nuclear weapons;
- LATENCY TRANSFORMATION of previously non-weapons technological and intellectual infrastructure to build, possess, integrate, maintain, and threaten to use nuclear weapons.

Nuclear-weapons proliferation can be thought of as an ensemble of covert acts (*e.g.*, NM theft, diversion) and overt acts (*e.g.*, "breakout" leading to the creation of a "ready threat"). Generally, Latency Transformation is related to the overt breakout phase, which can be

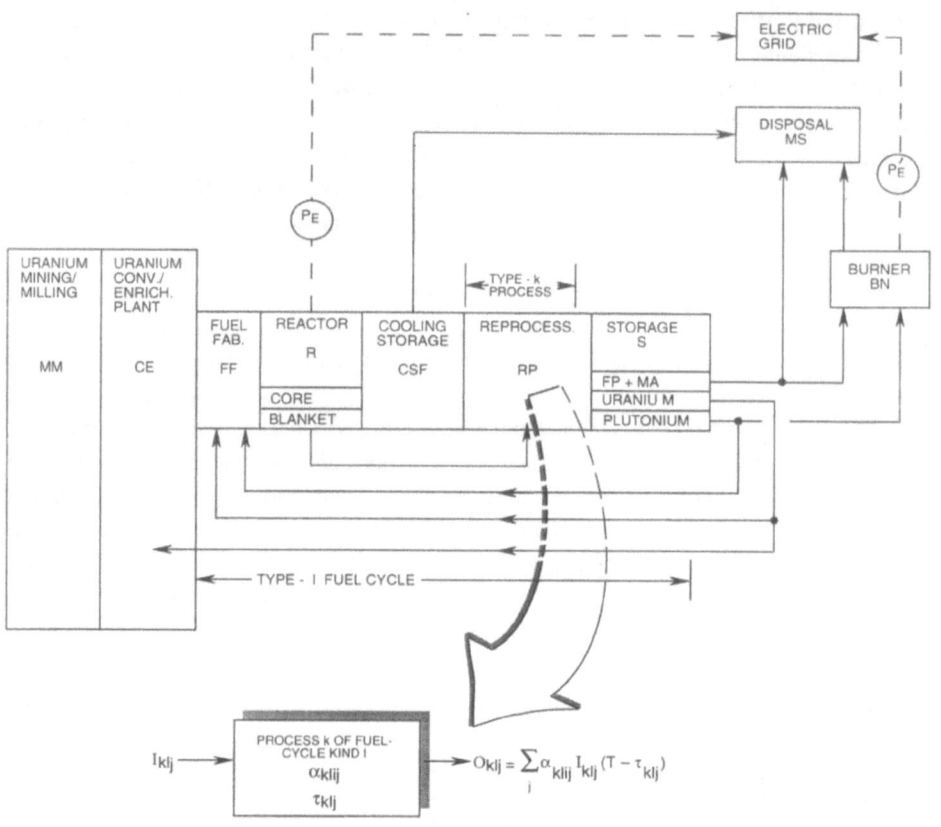

Figure 3. Condensation of the nuclear fuel cycle into a series of generic fuel-cycle building blocks for use in projecting (uranium) resource-constrained, multi-variable optimizations[28] of nuclear-energy mixes and nuclear-material flows. Also shown is a generalized process-flow diagram illustrating formulations used in Refs. 26 and adopted for the fuel-cycle analyses performed in the global E^3 model.

characterized by distinct processes that are subject to rate limitations, whereas the covert actions often can be characterized by threshold phenomena.

Figure 4 illustrates the funneling of these five drivers into the proliferation/breakout decision; eight specific NM safeguard conditions[37] are also indicated. If a "go" decision emerges from the "sum" of the five input drivers, the actions listed in the box on the right of Fig. 4 become rate determining insofar as the time at which a "ready threat" actually appears is concerned. The dynamics of this post-decision logic is illustrated on the second frame of Fig. 4. Many of the inputs listed on Fig. 4 that flow into the Incentives, Disincentives, and Latency Transformation decision elements are determined by economic, human-welfare, social/political, environmental, and energy-security issues, that in turn can provide either a negative or a positive feedback to the Materials Availability/Access driver in the proliferation decision. While many of these "variables" are subjective and driven by complex political-institutional-cultural forces that are beyond most simulations, the E^3 model can provide

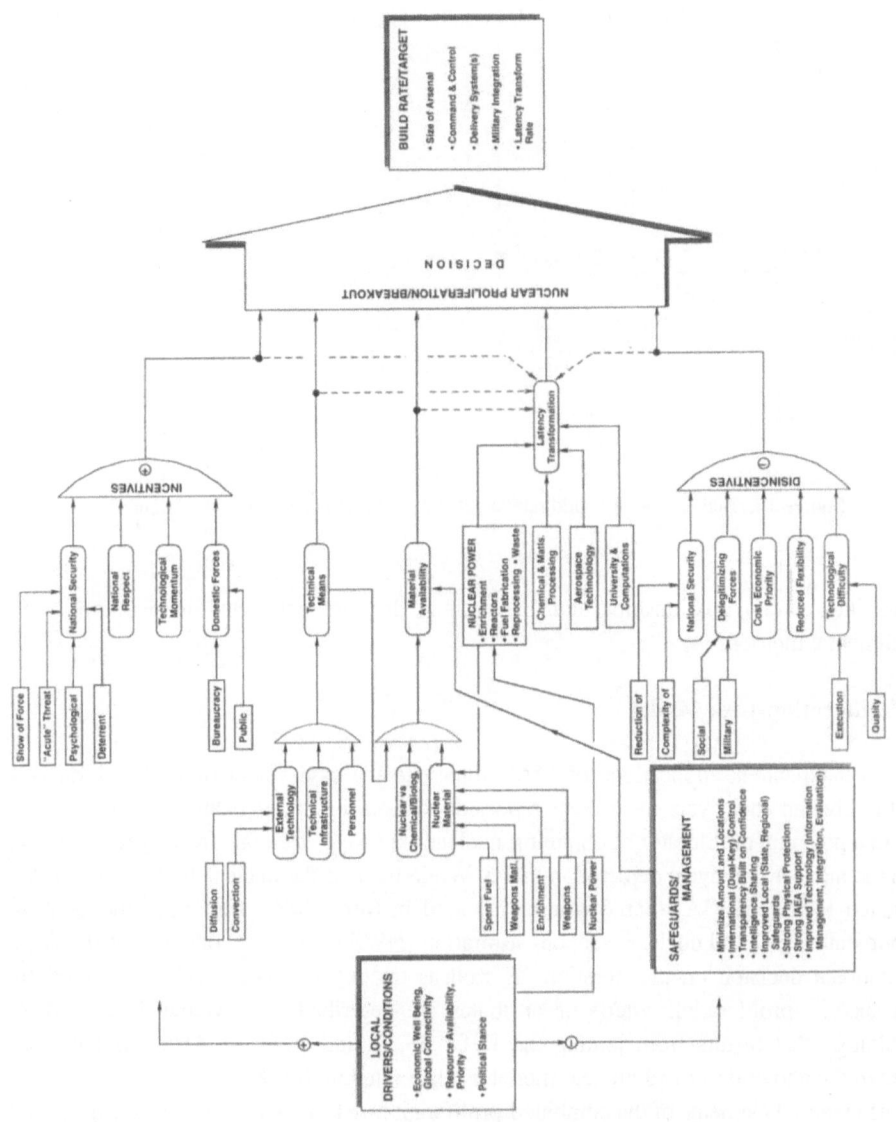

Figure 4A. Systems framework suggested for assessing proliferation risk. A: Overview.

209

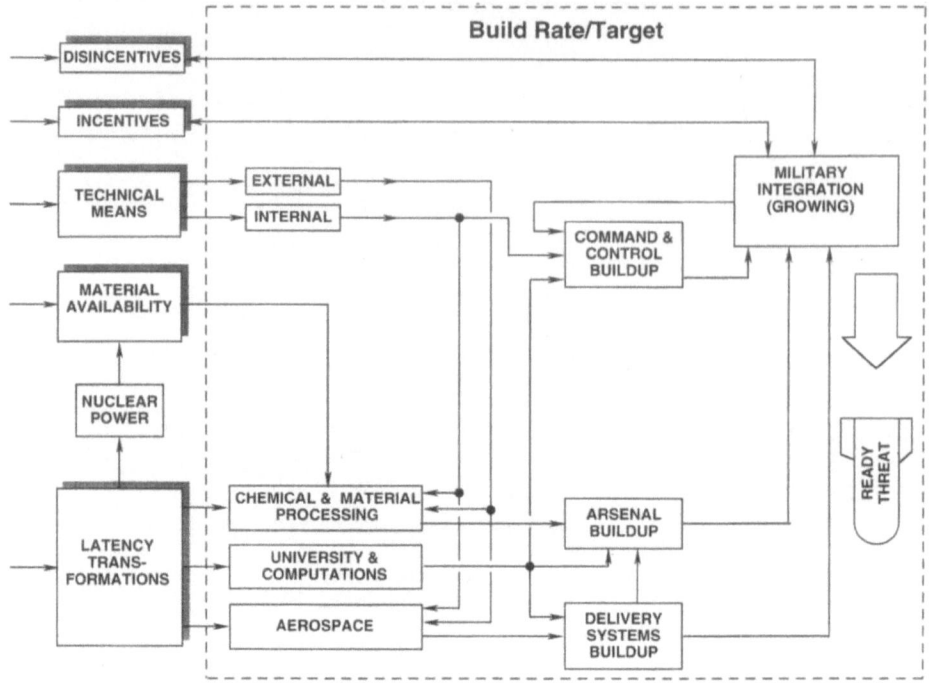

Figure 4B. Elaboration of "build rate/target" box indicated on part A of this figure.

insights into certain relationships and interactions that are connected to economic and environmental metrics.

Proliferation-Risk Model

Two independent applications of MAU theory to the assessment of proliferation risk from the civilian fuel cycle have been reported. References 38,39 examined the value or utility to a potential proliferator of obtaining nuclear-explosive materials from specific points within the nuclear fuel cycle depicted in Fig. 3. While treating the nuclear fuel cycle in more aggregated form, the MAU-based studies reported in Refs. 40,41 treat both the political environment (ENV) and nuclear-weapons aspiration (NWA) level that together set the stage for a national decision on proliferation, as well as treating in more detail the method by which specific proliferation criteria or attributes are described and evaluated. The MAU methodology that results from joining the Refs. 38,39 and Refs. 40,41 approaches, as applied to the above-described nuclear model, is elaborated in Ref. 21.

The essential elements of the combined proliferation-risk model are given in Fig. 5 and the appended footnotes.[21,40] The ENV and NWA parameters are specified for each global region as a function of time. The ENV and NWA parameters are used, along with attribute or criteria basis (normalization) parameters, to establish the shape of utility and subutility functions posited to describe each of five (proliferator-oriented) criteria[40]: Development Time (DT); Warning Period (WP); Inherent Technical Difficulty associated with Material Processing (ITD_{MP}); Inherent Technical Difficulty associated with NW fabrication (ITD_{NW});

and Cost (CST). Referring to Fig. 5, once ENV, NWA, and the state of sanctions (SANC) are specified for a given global region and time, and using the MOX/LWR and LWR/LMR mixes as a proxy for the fuel cycle FC_k, the $j = 5$ attributes are applied to each of $i = 4$ (HEU, SPU, MOX, and SFT) material streams. Plutonium undergoing fissioning in a reactor, RPU, is not included at this point in the proliferation-risk assessment, under the assumption that reactor plutonium actively undergoing fission is "safe and secure". Reference 21 summarizes details for each of the j criteria or attributes and each of the i material streams that comprise the core of the proliferation-risk model.

An important element in the estimation of each of the five (normalized) utility functions depicted on Fig. 5 (*i.e.*, u_{DT}, u_{WP}, $u_{ITD_{MP}}$, $u_{ITD_{NW}}$, and u_{CST}) is the assumption of separability (independence) of the plutonium inventory dependence from the dependence on the respective five utility criteria metrics (*i.e.*, x_{DT}, x_{WP}, $x_{ITD_{MP}}$, $x_{ITD_{NW}}$, and x_{CST}). Furthermore, the inventory-dependent part of each utility function is assumed[21] to approach unity exponentially at some threshold inventory for a given material stream i. The values of these inventory thresholds (above which the dependence of the utility function on plutonium inventory vanishes for that material stream), as well as the weights w_{ij} used to evaluate the the stream utility, $u_i = \Sigma_i w_{ij} u_{ij}$, impacts the utility for material stream i in each global region l at each time m, u_{ilm}, as well as affecting the associated Proliferation Risk Index, PRI.[21]

Using weights generated from pairwise comparison techniques[38,42], weighted utilities for each material stream are generated as a function of time for each global region, as described above. These material-stream utilities are then time-discounted and summed to give a Proliferation-Risk Index[28,39], PRI_{ilm}, for stream i, region l, and time m. This PRI value represents a geometrically weighted average of the particular LWR/LMR mix, as determined by the cost-minimized, market-penetration-rate-constrained value of the ratio LWR/LMR. Generally, the material stream with the maximum PRI is selected as the index to be monitored, $PRI_{lm} = MAX\{PRI_{ilm}\}$. Lastly, use once again of pairwise comparison techniques[42] to weigh the importance of region l compared to a reference region l′ in terms of importance of the respective value PRI_{lm} gives weights needed to generate a time-evolving global proliferation-risk index, PRI_m, relative to a reference region (*i.e.*, the U.S.). Hence, the time-dependent global PRI report here represents a highly aggregated metric that is sensitive to both the (plutonium) inventory dependence of the utility function, as well as the serial weighting process used.[21]

RESULTS

Overview

A sampling of the typical results from the modified ERB model given in Ref. 22 are reported herein. Table I summarizes an ensemble of six cases that is centered on "business-as-usual" (BAU) conditions (Case A) and have been selected primarily for illustrative purposes. Except for the modifications described above, the BAU Case-A parameters are those in the "as-received" version of the ERB model[12,25]. Table II lists key parameters used in the nuclear model that runs "under" the ERB model. Most of these parameters remained fixed for the BAU and the five perturbation cases described in Table II, and differ somewhat from those reported in Ref. 20. None of the extensive parameter input required to evaluate the proliferation-risk model are given here[21]. While both the nuclear fuel-cycle/material-flow

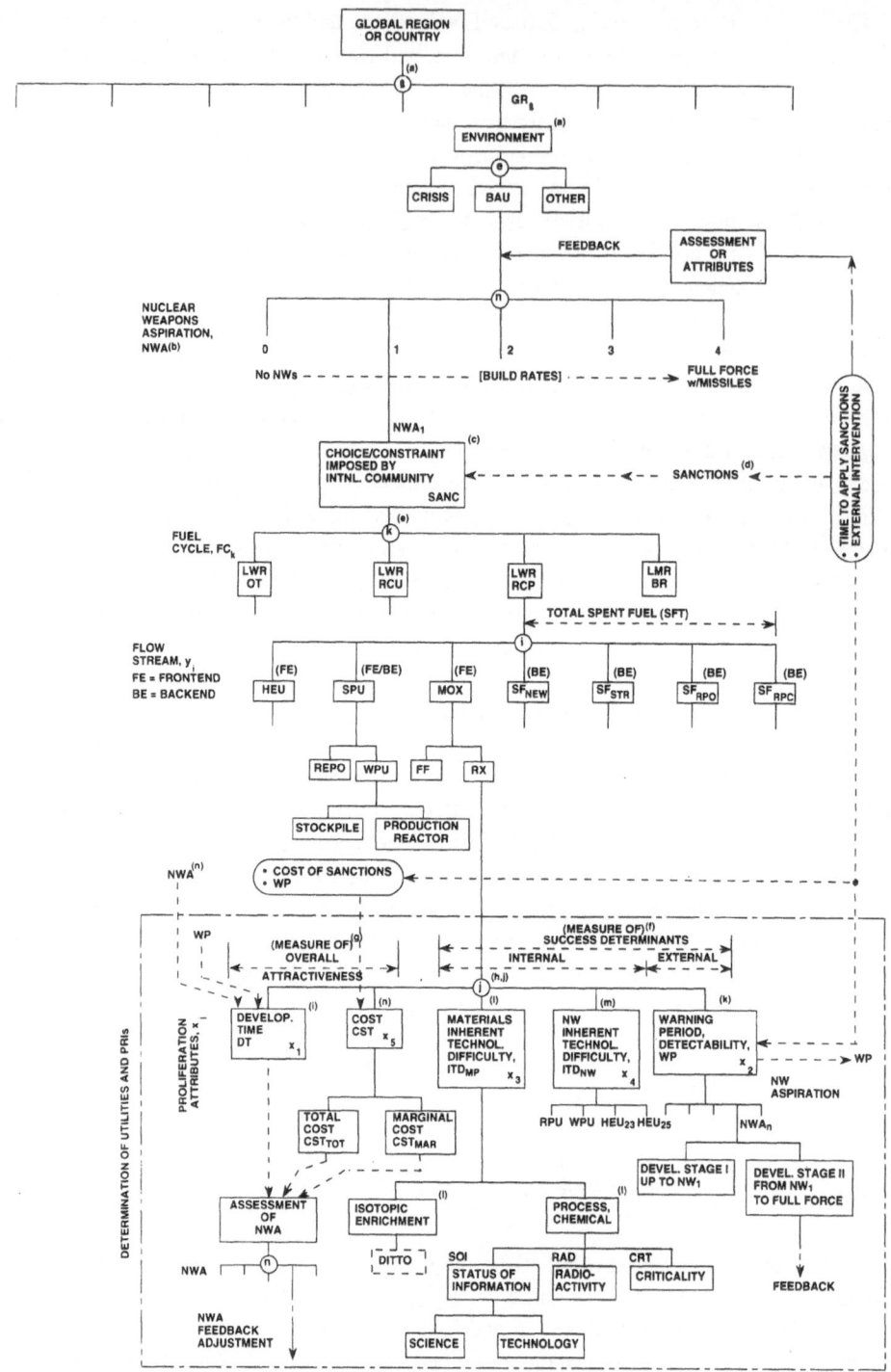

Figure 5. Logical sequence diagram depicting algorithm for assessing proliferation risk as determined by combining the MAU-based approaches reported in Refs. 38,39, and Refs. 40,41; refer to the Nomenclature for definition of symbols/acronyms. Footnotes for elaboration of Fig. 5 largely from Ref. 40.

(a)   The choice of country/region (l) sets characteristics, preferences priorities, *etc.* through which the political environment (ENV) and level of nuclear-weapons aspiration (NWA) can be specified; both

ENV and NWA for a given region l may (in future work) change with time, as well as with similar conditions existing in other countries/regions.

(b) The levels of NWA = (0,4) corresponds in rank order to: a) no NWs (0); one crude non-deliverable device (1); c) 10 deliverable NWs per year (2); d) 100 deliverable NWs per year (3); and e) >1000 deliverable NWs per year (4); depending on the rank order, a value is assigned to NWA_{lm} that is used to modify the risk-aversion factors (RAF_i or RAF_j, which are shape factors for the specific utility functions) for that region at that time.

(c) The choices/constraints imposed through sanctions ($SANC_{lm}$) by the international community can impact: a) the kind of fuel cycle available or allowed; b) the kind of facilities allowed within the country/region; and c) various international agreements on what constitrutes the illegal use of facilities and the sanctions related thereto.

(d) Sanctions $SANC_{lm}$: political pressure; curtailment/cancellation of credits, technology exchanges, agreements of cooperation, particularly with respect to nuclear assistance; variations between "light" sanctions to "heavy" sanctions (including preemptive strikes).

(e) The parameter k defines a given (total) fuel-cycle system which, for a given country/region, includes technological characteristics, institutional constraints, and sanctions to be imposed in case of violation and discovery thereof; in the context of the present study, the parameter k is used to differentiate between LWRs and LMRs and any conditions applied temporally or regionally to exclude or constrain technologies associated with either or both.

(f) Attributes j = 2, 3, and 4 (*i.e.*, WP, ITD_{MP}, and ITD_{NW}) contain all the elements that affect the probability that the (proliferation) effort will be successful for a given set (ENV_{lm}, NWA_{lm}, FC_k, and y_i); these attributes are further divided into "external" and "internal" (to the proliferation effort) classes, where again proliferation $\equiv \{(theft,diversion),breakout\}$ is used in a limiting sense of diversion from the civilian fuel cycle, and these attributes are evaluated and interpreted in that context..

(g) Attributes j = 1 and 5 (DT and CST) contain all the elements that make one pathway more desirable than another, given the successful completion of the effort[(f)].

(h) The following characteristics are used in selecting these j = 5 attributes: complete; operational; non-redundant; and minimum size (*e.g.*, number and complexity).

(i) The following diagram depicts the four primary time lines that determine the NW Development Time, DT:

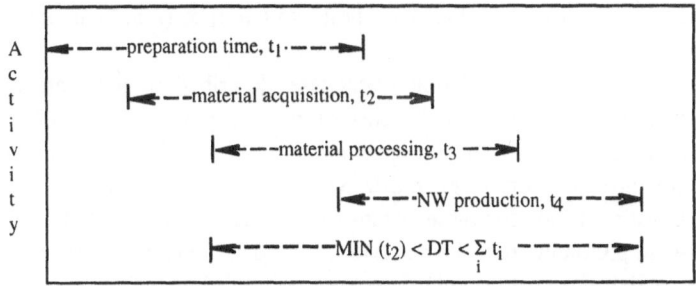

Time

Costs (CST, x_5) and probabilities for detection or Warning Period (WP, x_2) can be associated with phases of NW development. A simpler (single phase) approach is taken in the present model for parametrically assessing risk in the context of the ERB global E^3 model[8,20], but future work will extend the model to include the above-described development phases.

(j) The proliferation pathway/process/procedure is divided into two Phases and two Modes:

Phases:

- preparation/prediversion: R&D, facilities design and construction, but no diversion of material from the commercial operations;
- diversion: nuclear fuels, materials being diverted.

Modes:

- covert: slower progress (higher DT), reduced probability of detection (reduced WP);
- overt: faster progress (reduced DT), greater activity and higher probability of detection (increased WP).

The combination of Phases and Modes gives 4 "top-level" pathways. Provisions are made in the model for incorporating these features, however, actual specification and understanding of the means to do so are in a preliminary/elementary stage.

(k) The Warning Period, WP, is measured in terms of the fraction of the NW development/production task[(i)] remaining to be completed at the time of detection. As discussed in Ref. 21, this fraction,

$(1 - t_D/DT)$, where t_D is the time of detection, can be expressed in terms of a growing probability of detection (*e.g.*, a "failure" rate in the parlance of reliability theory) and an NW production function, $P(t)$, giving the fraction of the NW development/production activity remaining at time t. Important tradeoffs exist with most of the other (assumed independent) attributes used to assess the over all proliferation risk.

(l)   The inherent technical difficulty associated with material processing, ITD_{MP}, is divided into isotopic-enrichment and chemical processing operations, each being interpreted in terms of information availability (both technological and scientific), radiation hazard, and criticality hazard; all evaluations must be based on the (evolving) scientific, technological, and industrial capability and infrastructure of a given country/region.

(m) The inherent technical difficulty for actual NW fabrication, ITD_{NW}, is also dependent on the (evolving) scientific, technological, and industrial capability and infrastructure of a given country/region; this attribute, however, is measured only in terms of the isotopic composition of the nuclear explosive being assembled, which for plutonium is measured in terms of integrated fuel burnup, BU(GWtd/tonne); HEU is given the highest utility[40], although clandestine uranium enrichment in competition with sources from the civilian fuel cycle is not included in the results reported herein.

(n)   The cost attribute, CST, is divided into total cost, CST_{TOT}, and marginal cost, CST_{MAR}. Total cost is related to proliferation pathways base on a "greenfield" construction endeavor and includes direct capital, O&M, personnel, and capital-service charges that are integrated over relevant periods of time and discounted to the present. Marginal costs include the costs of replacing lost power (if nuclear material is taken from the commercial sector, and donor power-plant availability is decreased), as well as the cost of sanctions; the cost of sanctions is strongly coupled to the WP and DT attributes, the NW-development/production technologies selected, and the probabilities that sanctions will be enforced at a given level once the proliferation action is detected.

and proliferation risk models have been developed to express broad regional and temporal variations, most of the illustrative results reported here model a world that, aside from energy-demand-driven NM inventories and the final weights used to generate the global PRI, is relatively uniform from the viewpoint of proliferation risk (*e.g.*, parameters listed on Fig. 5).

The essential elements differentiating the six cases described in Table I are: a) the degree of plutonium recycle (to LWRs); b) the cost of nuclear power; c) the LWR/LMR mix (controlled by relative costs); and d) the imposition of a fossil-fuel carbon tax. Control of the NE market share through costs (*e.g.*, either NE capital cost or tax-enhanced fossil-fuel prices) is a motivating factor for these parametric choices. The BAU (base) Case A is characterized by no plutonium recycle, no LMRs, and no carbon taxes. Case B examines the impact of increased LWR costs, while (exogenously) disallowing an economically viable LMR option. The impact of increased fossil-fuel charges imposed by a (naively) simple carbon tax is investigated in Case C; this carbon tax will ultimately serve as a proxy against which the costs of sustainable technologies (*e.g.*, like NE or biomass) that reduce global GHG emissions by a given amount must be compared on an integrated, present-value basis[12]. Introduction of LMRs (of reduced cost compared to the BAU base case) is considered in Case D. Both Cases E and F examine the impact of plutonium recycle to LWRs without carbon taxes imposed on the use of fossil fuels and without the introduction of LMRs; these two plutonium recycle cases differ in whether or not a regional dependence is implemented on the rate of and level at which is plutonium recycled.

The results described in the following section represent a sampling from more extensive results collected in Ref. 22. The primary motivation here is: a) to describe the BAU case in comparison with results derived from other studies; b) to model the accumulation of plutonium under a limited range of fuel-cycle scenarios; and c) to illustrate a preliminary relationship between impacts associated with NM inventories (proliferation) and GHG accumulations (global climate change).

Sample Results

Business-As-Usual (BAU) Base Case

The long-term energy scenario computed by the ERB model is driven by exogenous inputs of (regional) population and (base) GNP growths, along with assumptions about regional technology improvements, labor force, and tax and tariff rates.[8] The base GNP input is modified to reflect the evolution of world fossil-fuel prices needed to clear markets in each of the nine 15-year time steps that take the computation out to the year 2095 (starting from 1975). Figure 6 gives the population-growth driver, along key aggregated economic and energy parameters that result for the BAU Case A. The population growth assumed for the BAU Case A stabilizes at ~10.5 billion persons, and price-adjusted GNP growth rate decreases from 2.6 %/yr over the period 1990-2005 to ~1.7-1.8 %/yr in the latter part of the 21st century. The *per-capita* energy consumption is relatively flat out to the year 2030, and increases slightly thereafter. Likewise, the *per-capita* GNP shows a steady increase from the present ~4 k$/yr to 20 k$/yr by the year 2100, whereas the decrease in global energy intensity shows an improvement from ≥ 10 MJ/$ to ≤ 6 MJ/$. Regional and temporal breakdowns of these demographic and economic parameters are provided by the ERB model results[22].

Primary Energy Usage. Aggregation of the nine global regions into industrialized and developing sectors (*e.g.*, referring to the Nomenclature, Industrialized Countries = USA + OECD-E + OECD-P + FSU/EEU; Developing Countries = CHINA+ + ME + AFR + LA + SEA) leads to Fig. 7; generally, the cross-over in total (annual) energy demand is ~10-15 years later than suggested from other studies[43]. The breakdown into primary energy

Table I. Summary of Case Characteristics/Parameters

Case	LWR Plutonium Recycle	LWR Unit Cost UTC_{LWR} ($/We)	Relative LMR Cost UTC_{LMR}/UTC_{LW} R	Carbon Tax
A	No	Nominal[a]	High[b]	No
B	No	High[c]	High	No
C	No	Nominal	High	High[d]
D	No	Nominal	Nominal[e]	High[d]
E	Yes[f]	Nominal	High	No
F	Yes[g]	Nominal	High	No

(a) $UTC_{LWR} = 2.0$ $/We

(b) $UTC_{LMR} = 2.0 \times UTC_{LWR}$

(c) 4.0 $/We

(d) 40 $/tonneC/15yr, starting in 2005

(e) $UTC_{LMR} = 1.3 \times UTC_{LWR}$

(f) all regions achieve a 15% MOX core fraction in a 40-yr exponentiation time; some regions experience a plutonium deficit before ~2100, but global plutonium in spent fuel is constrained to remain positive.

(g) region-dependent final MOX fraction and exponentiation time; all regions do not sustain a plutonium deficit before ~2100.

Table II. Summary of Key Input (Fixed or Computed) to Nuclear Fuel-Cycle Model

INITIAL PARAMETERS[a]	
Accumulated uranium mined by 1990, $U_o(kg)/10^8$	2.00
Installed nuclear capacity as of 1990, $P_{Eo}(MWe)/10^5$	3.40
Accumulated spent fuel as of 1990, $M_{SFo}(kg)/10^7$	5.89
Accumulated MOX fuel as of 1990, $M_{MOX}(kg)/10^6$	3.00
Initial plutonium in reactors, $M_{Puo}(kg)/10^5$	2.56

LWR REACTOR PARAMETERS	
Fuel burnup, BU(MWd/kgHM)	40.0
Fuel replacement or life time, $\tau_R(yr)$	3.0
Specific inventory, SI(kg/MWt)	26.7
Fraction of all actinides that are minor, f_{MA}	0.0
Fraction of all plutonium that is fissionable, f_{Puf}	0.60
Thermal-to-electric conversion, η_{TH}	0.325
Plant availability, p_f	0.70
Fraction of fissions from ^{235}U, f_{25}	0.60
Engineering gain, Q_E[b]	25.

ENRICHMENT AND (LWR) REACTOR CONCENTRATIONS	
Plutonium concentration in spent fuel, f_{Pu}	0.0090
Effective breeding ratio for OT→MOX, BR_o	0.3018
Weight fraction ^{235}U in ER product stream, x_p	0.0300
Weight fraction ^{235}U in ER feed stream, x_f	0.0071
Weight fraction ^{235}U in ER tails stream, x_t	0.0020
Weight fraction plutonium in MOX, x_{Pu}	0.0400
Weight fraction ^{235}U in RU stream, x_d	0.0121
Total burnup fraction for LWR, x_{BU}	0.0298
^{235}U burnup fraction, x_{BU25}	0.0179
Uranium ore grade (weight fraction), $x_{ORE}/10^{-6}$	5.00

LMR PARAMETERS	
Thermal-conversion efficiency for LMR, η_{TH}	0.40
Plant availability, p_f	0.70
Burnup for LMR, BU(MWd/kg)	80.0
Breeding ratio for LMR, BR	1.00
Fuel concentration in LMR, x_{Pu}	0.10
Specific inventory for LMR, SI(kg/MWt)	67.6
Total burnup fraction for LMR, x_{BU}	0.0596
Simple doubling time for LMR, DT(yr)	∞
Engineering gain, Q_E	25.
Market penetration time constant, $\lambda_c(1/yr)$	0.169

RECYCLE PARAMETERS	
Initial fraction of load supplied by MOX, f_{MOXo}	0.0010
Final fraction of load supplied by MOX, f_{MOXf}	0.0 → 0.30
Half-time for $f_{MOXo} \rightarrow f_{MOXf}$, $T_{MOX}(yr)$	30. → 40.

Initial separated plutonium inventory, $M_{SPU0}(kg)/10^5$	1.20
Time when $f_{MOX0} \rightarrow f_{MOXf}$ rampup starts, $t_{MOX0}(yr)$	1990
MOX core burnup, $BU_{MOX}(MWd/kg)$	40.0
Hold-up time for LWR reprocessing, $\tau_{RP}(yr)$	1.0
Hold-up time for LWR fuel fabrication, $\tau_{FF}(yr)$	0.5
Hold-up time for LMR reprocessing, $\tau_{RP}(yr)$	0.3
Hold-up time for LMR fuel fabrication, $\tau_{FF}(yr)$	0.3

COSTING PARAMETERS

Fitting constant for uranium ore cost, U_1	0.00031
Fitting constant for uranium ore cost, v	0.6300
Unit cost of uranium ore in 1990, $UC_{MM0}(\$/kgU)$	52.6
Unit cost of uranium conversion, $UC_{CV}(\$/kgU)$	5.0
Unit cost of uranium separative work, $UC_{SW}(\$/kg\ SW)$	100.0
Unit cost of uranium fuel fabrication, $UC_{FF}(\$/kgU)$	200.0
Unit cost of MOX fuel fabrication, $UC_{FF}(\$/kgU)$	400.0
Unit cost of spent fuel storage, $UC_{SF}(\$/kg/yr)$	10.0
Unit cost of fission-product storage, $UC_{FP}(\$/kg/yr)$	10.0
Unit cost of SF/FP transport, $UC_{TR}(\$/kgU)$	0.0
Unit cost of reprocessing LWR, $UC_{RP}(\$/kgHM)$	1000.
Unit cost of reprocessing LMR, $UC_{RP}(\$/kgHM)$	1500.
Unit total cost for LWR, UTC_{LWR} ($\$/We$)	$1.25 \rightarrow 2.0^{(c)}$
Unit total cost factor, $f_{UTC} = (UTC_{LMR} - UTC_{LWR})/UTC_{LWR}$	1.0
Unit total cost for LMR, $UTC_{LMR}(\$/We)$	4.0[d]
Fixed charge rate for LWR, $FCR_{LWR}(1/yr)$	0.09
O&M charges as fraction total capital for LWR, $f_{OM}^{LWR}(1/yr)$	0.02
Fixed charge rate for LMR, $FCR_{LMR}(1/yr)$	0.09
O&M charges as fraction total capital for LMR, $f_{OM}^{LMR}(1/yr)$	0.02
Unit cost(value)[e] of plutonium, $UC_{PU}(\$/kgPu)/10^4$	1.85
Unit cost(value)[e] of MOX, $UC_{MOX}(\$/kgPu)$	738.

[a] global values.

[b] ratio of total electric power to power used on site; the net plant efficiency is $\eta_P = \eta_{TH}(1 - 1/Q_E)$.

[c] varied with time.

[d] value selected to assure LMR does not compete with LWR under BAU conditions.

[e] value based on saved enrichment costs, accounting for reprocessing and (increased) fuel-fabrication costs[28].

categories of this global demand is shown in Fig. 8 for the Case-A conditions. The time dependence of the global mix of the six primary energy sources [e.g., oil, gas, solids (coal + biomass), nuclear, hydroelectric, and solar-electric (solar-thermal energy is described in theERB model as a conservation measure)] generally reflects a large unit costs used for solar-electric and biomass, which results in the relatively small shares [only ~4.3% of the solid primary energy consumption (labeled in Fig. 8 as "coal") is attributed to biomass in 1990, which grows to 9.0% by 2005, 9.3% by 2095]. Generally, the BAU Case A suggests a growth of 1.2-1.3 %/yr in primary energy usage after the year 2000, compared to 1.9 %/yr (1975-1990), 2.1 %/yr (2035-2050) and 1.8%/yr (2080-2095) for the exogenous global

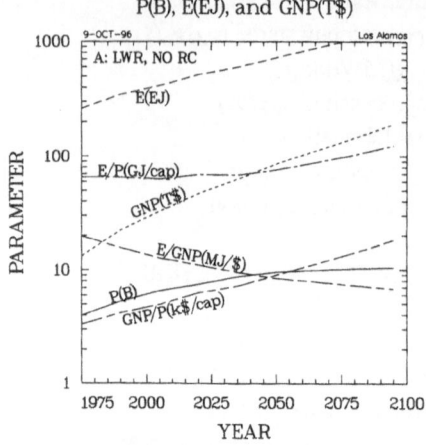

Figure 6. Time evolution of global population, P (an exogenous input); total primary energy demand, E; Gross National Product, GNP (baseline input, modified by evolving energy prices); primary energy per GNP, E/GNP; primary energy per capita, E/P; and GNP *per- capita*, GNP/P; for the BAU Case A (Table I).

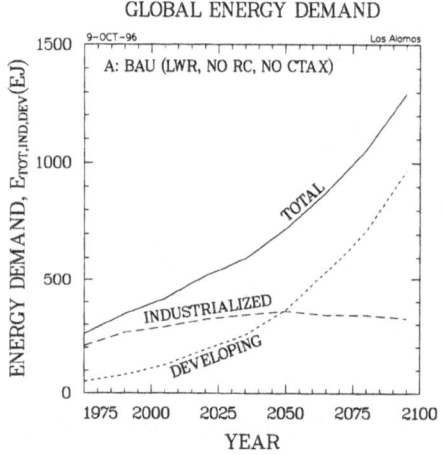

Figure 7. Base-case BAU total primary energy demand as a function of time, showing a breakdown into industrialized (USA + OECD-E + OECD-P + FSU/EEU) and developing (CHINA[+] + ME + AFR + LA + SEA) global sectors; refer to Nomenclature for definitions.

population growth. The growth in world GNP decreases from 3.5 %/yr (1975-1990), to 2.1%/yr (2035-2050), and to 1.8%/yr (2080-2095).

Comparisons. Figure 9 gives a collection of recent prognostications, including BAU, high-efficiency, and/or sustainable scenarios, against which the BAU Case A is compared. While the range of global demand projections is large, with some of this range being created

CUMULATIVE GLOBAL ENERGY MIX

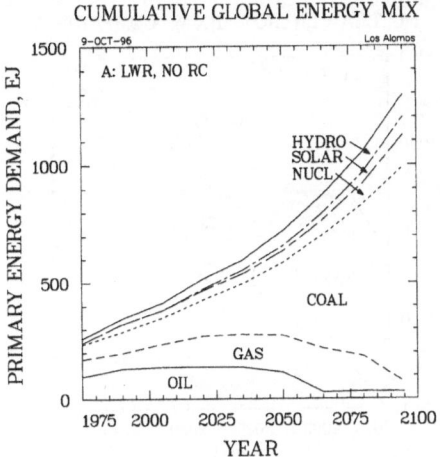

Figure 8. Time dependence of global demands for primary energies for the BAU Case A.

GLOBAL ENERGY DEMAND w/ COMPARISON

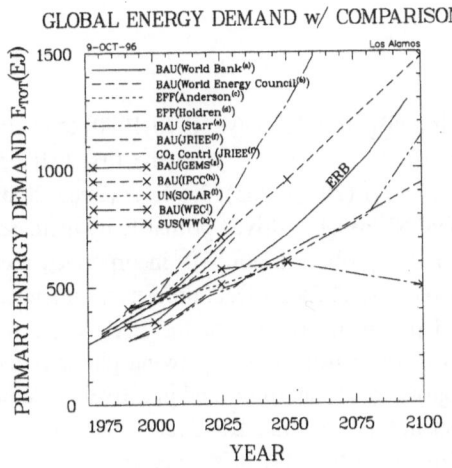

Figure 9. Comparison of BAU (Case A) primary-energy demand scenario generated from the ERB model with other studies:

(a) Refs. 44,45; (b) Ref. 46; (c) Refs. 44,47; (d) Ref. 44;
(e) Ref. 48; (f) Ref. 48; (g) Ref. 50; (h) Refs. 51,52;
(i) Ref. 52,53,58; (j) Ref. 46,53; (k) Ref. 52.

by the inclusion or exclusion of indigenous fuel sources[44], the ERB Case A is in the mid-to-upper range of other projections, while indicating a later cross-over of industrialized- *versus* developing-country demand.

Nuclear Energy. The temporal and regional dependence of nuclear energy is shown in Fig. 10, which gives annual nuclear-electric consumption rather than capacity. The nominal

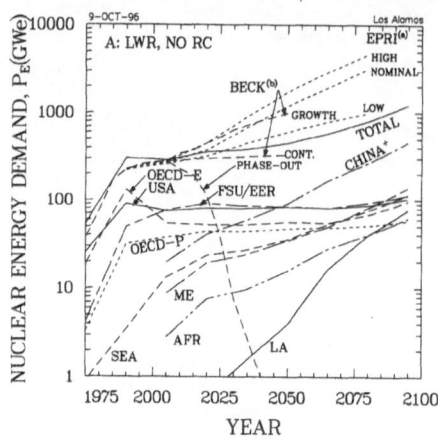

Figure 10. Time and regional dependencies of nuclear-energy demand for the BAU Case A (annual energy generation rather than capacity is reported), including recent projections/scenarios.

(a) Ref. 54; (b) Ref. 55.

annual growth decreases initially from 9.3 %/yr (1975-1990) to negligible values around the year 2000, but then increases to ~1.1-1.4 %/yr in the first half of the 21st century, with subsequent increases of 1.7 %/yr (2050-2065), 2.1 %/yr (2065-2080), and 2.7 %/yr (2080-2095); this growth in global NE usage is driven primarily by increased use in the developing regions (mainly China) of the globe. Figure 10 incorporates recent NE scenarios[54,55], which indicate that the BAU Case-A ERB model projects a somewhat lower NE demand.

The growth in NE depicted in Fig. 10 for the BAU Case A (LWRs, no plutonium recycle, no carbon taxes) is accompanied by a growing global inventory of plutonium that resides primarily in a range of spent-fuel forms and in actively fissioning reactor inventories. The total [accumulated (spent fuel), separated, and reactor] plutonium inventories are shown as a function of time and region in Fig. 11. It is noted from Fig. 11 that the majority of the civilian plutonium inventory, for historical reasons, resides in the U.S., other OECD countries, and the FSU well into the mid-21st century. The proliferation risk, as measured by either the material-stream maximum utility or the proliferation-risk index (which is the discounted sum of utilities using a 2 %/yr discount rate[39]), for the BAU case for a given region at a given time, is driven primarily by the inventories and related utilities associated with the spent-fuel material stream. Depending on fuel-cycle scenario and NE growth rates, the fuel-cycle stream that is contributing to the PRI evolves and shifts in time[21,22].

Proliferation Risk. The regional dependence of PRI is dominated by the developed countries, which have the larger inventories of spent fuel. The parameters used to generate these proliferation-risk metrics are summarized in the Ref.-21 model description. As noted previously, proliferation can be viewed as a sum or ensemble of covert acts (*e.g.*, thefts, diversions) and overt acts (*e.g.*, breakout). At the present level of fidelity, the model does

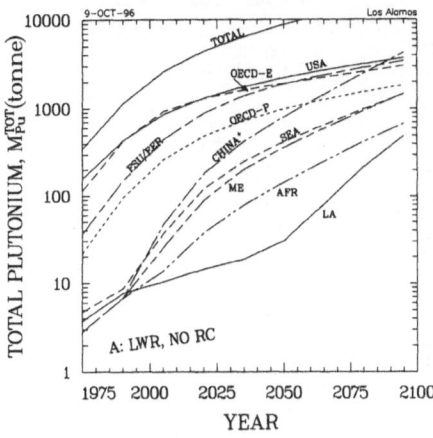

Figure 11. Time and regional dependencies of total plutonium inventories for BAU Case A; since little or no plutonium recycle is allowed for this case, separated plutonium inventories (in processing, in a range of separated forms, and in fuel fabrication as MOX) are negligible (re: Fig. 14).

not distinguish between "proliferation", "diversion", or "breakout". Furthermore, the degree to which the risk associated with either depends on the actual inventories in a given material stream remains uncertain, but the threshold inventories used to generate the results reported here[21] are set at relatively high values. Generally, the global proliferation risk must be determined using a weighting of the regional PRIs at a given time. The high inventory-driven PRI for the U.S.A., for instance, from the perspective of the United States would be given a low weighting relative to other global regions having lower PRIs. Using such a weighting scheme[21], the time-evolution of the global PRI and associated global utility depicted in Fig. 12 results. The magnitude of global PRI or utility must ultimately be related to a specific (global) consequence, but on a relative basis the global PRI presently serves as a proxy to be correlated with other global risks associated with energy use (*e.g.*, GHG emissions and accumulations) or non-use (*e.g.*, stunted improvements in human welfare, as measured crudely by *per-capita* GNP). As for the PRI metric, the consequence of GHG accumulations or reduced human welfare remain to be quantified in terms of specific and comparable consequences.

Carbon Dioxide Emissions. The global and regional emissions of carbon dioxide for the BAU Case A are shown as a function of time in Fig. 13. Similar estimates for N_2O and CH_4 are available from the ERB model. The dominance of China in contributing CO_2 to the atmosphere for the Case-A conditions occurs sometime after the year 2025.[22] The relatively long atmospheric residence time for CO_2 indicates that most of the integrated emissions depicted in Fig. 13 will be retained in the atmosphere on the time scale of this computation; the atmosphere presently contains a total gaseous carbon mass of $M_{CO_2,1990} = 740$ Gtonne[56]. In this model, the ratio of the integrated CO_2 emissions measured from the ERB base year (1975), M_{CO_2}, to $M_{CO_2,1990}$ is used as an index of GHG risk, and is correlated

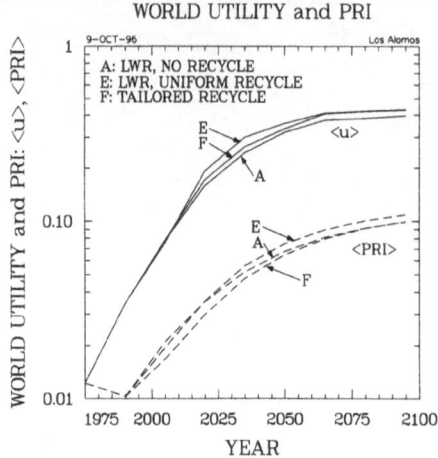

Figure 12. Time dependencies and comparisons of global proliferation-risk utility and index for range of recycle case embodied in Case A (no recycle), E (regionally uniform recycle), and F (regionally tailored recycle).

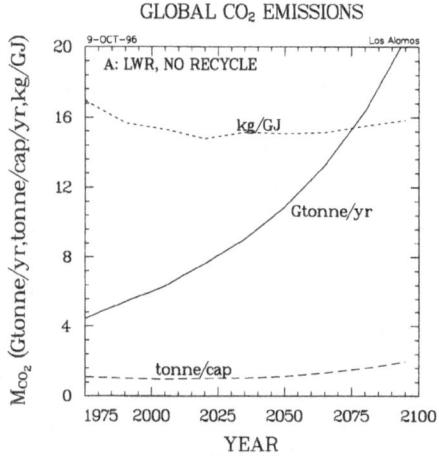

Figure 13. Global carbon emissions in the form of carbon dioxide for the BAU Case A.

with PRI; this ratio is defined as the greenhouse gas parameter, GHP. While this correlation indicates the expected increase in risks associated with proliferation and GHG emission with growths in population and the economic welfare of those populations, mapping this relationship into a "consequence space" through some yet-to-be-determined transformationcould change the form (*e.g.*, slope and curvature) of this relationship. Resolution of such effects, however, are outside the capabilities of the present model. The primary value of the PRI *versus* GHP = $M_{CO_2} / M_{CO_2,1990}$ relationship, at the present level of analysis, is as an indicator of relative shifts and changes when other scenarios (*e.g.*, Table I) are considered.

Plutonium Recycle

For the BAU Case A a minor amount of recycle of plutonium is allowed, and the separated plutonium in either reprocessing or in the form for MOX is small. Two cases involving plutonium recycle to LWRs are indicated on Table I: a) Case E allocates a rate and magnitude of plutonium recycle uniformly over the nine global regions; and b) Case F adjusts the MOX-recycle exponentiation time and final (region-averaged) MOX core fraction on a regional basis in a way that assures each regional inventory of accumulated (spent-fuel) plutonium remains positive. Without this local "tailoring" of MOX recycle fraction, local plutonium demands must be satisfied from plutonium excesses in other global regions; details of this "plutonium trade" on a world market (and the proliferation risks related thereto) remain for future elaboration.

The global inventories of plutonium in the accumulated (spent-fuel), reactor, and separated (reprocessing and fuel-fabrication) forms are shown as a function of time for the BAU Case A (no plutonium recycle) and the regionally uniform recycle Case E [40-yr exponentiation time to a final (average) MOX core fraction of 0.15; MOX core-volume fractions much above this value result in insufficient (spent-fuel) plutonium over the (~2100) computational period, for the simplified recycle conditions assumed] in Fig. 14. The decrease in total global plutonium inventory results from a balance in decreased plutonium in spent fuel *versus* increases in reactor and separated (reprocessing and fuel fabrication) plutonium. The small amount of separated plutonium for the BAU Case A reflects the small but non-zero value used for the initial MOX core fraction. Figure 15 gives the regional distribution of separated plutonium as a function of time for Case-E conditions.

With the non-zero inventories of separated [*i.e.,* reprocessing (SPU) and fuel fabrication (MOX)] plutonium in Cases E or F, the utilities related to proliferation evolve in time with a more varied dependence on material-stream inventories. For the inventory dependencies used[21], a shift in the material stream having the maximum utility from accumulated (SFT) to reprocessing (SPU), and eventually ($\geq$ 2100) to fuel-fabrication (MOX) material streams occurs. For the 2-%/yr discounting of proliferation utility used, however, the SFT material stream dominates the PRI over the time period and for the region (USA) considered. For this reason, as well as the closeness of the stream utilities in the out years, the specific weighting used to generate global proliferation utilities and PRIs result in little difference between the Case A, E, and F plutonium recycle strategies, as is indicated on Fig. 12. It should be noted that the approach used in the model for plutonium utilization is not particularly innovative or aggressive, and incorporation of more detail into the plutonium inventory/flow model will allow more aggressive approaches/scenarios to be examined.

Carbon Tax

The economically optimal implementation of energy taxes in general, and carbon taxes in particular, is complex and not necessarily unique. The Case C (Table I) applies a carbon tax uniformly to each global region. This taxation algorithm starts in the year 2005 and uniformly applies a fossil-fuel charge that increases at a rate of 40 $/tonneC per 15-year computational time interval. The ratio of carbon emitted per unit energy generated varies for a given fossil fuel[8,12]. The illustrative example reported here ignores important issues related to[57]: a) revenue recycle (*e.g.,* ways in which tax revenues are reinjected into economies to

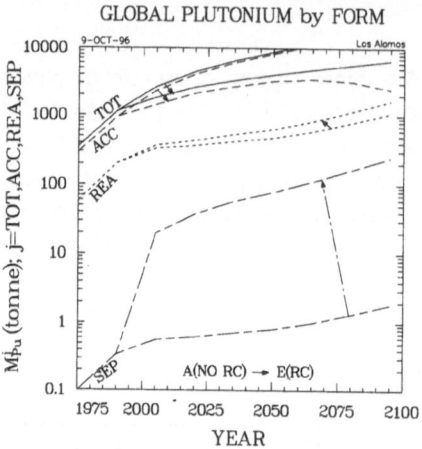

Figure 14. Time dependencies of global plutonium inventories (TOT ⇒ total; ACC ⇒ accumulated in all spent-fuel forms; REA ⇒ active reactor inventories; SEP ⇒ separated in both reprocessing and fuel-fabrication facilities), showing impact of LWR recycle applied uniformly across all global regions for a final MOX core fraction $f_{MOXf} = 0.15$ and the exponentiation time $\tau_{MOX} = 40$ yr (Case E).

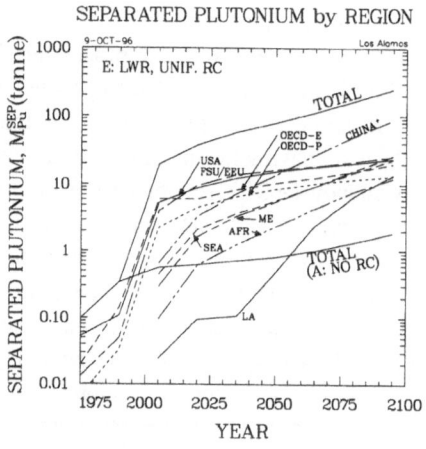

Figure 15. Time and regional dependencies of separated plutonium material streams for Case E (LWR, regionally uniform plutonium recycle).

reduce other distortionary taxes like personal income, corporate, or payroll taxes); b) subsidy removal (which in 1990 amounted to 215 B$, and the elimination of which would reduce carbon emission by 20% in subsidized countries and worldwide by 7%[56,57]); c) and the trading of emission "rights" in a way that would level the marginal cost of GHG reduction across the globe. The example reported here illustrates the impact of a frontal strategy to reduce GHG emissions on market shares of NE and other other energy forms, as well as aggregated economic impacts.

The imposition of a (fossil-fuel-specific) carbon tax and the resulting increase in fossil-fuel prices has three effects: a) decrease the used of fossil fuels; b) increase the market share of more expensive, reduced-carbon or non-fossil energy sources; and c) decrease the regional and global GNP. A comparison[22] of the counterpart to Fig. 8 with carbon taxes applied shows both a decrease in total energy consumption, a decrease in the most egregious carbon-emitting fuel (coal), an increased use of more-expensive but reduced carbon-emitting oil and gas, and increases in nuclear and the more-expensive solar energy sources.

The use of more-expensive energy by the (heavily) carbon-taxed world results in a decrease in the GNP, energy intensity (MJ/$), and GNP *per capita*. Slight shifts in the global distributions or "concentrations" of these parameters also result.[22] The impact of a range of carbon-tax rates on the CO_2 emission rate is shown on Fig. 16, which includes a number of CO_2-emission scenarios[51-53,58] for comparison. A plot of GNP reduction *versus* CO_2 reduction[22] indicates an increasing sensitivity of GNP as the carbon tax rate is increased to induce greater reductions in CO_2 emissions. The slope or "elasticity" is 0.89 for reductions in CO_2 emission rates below ~20% and increases to 1.53 for the higher rates of CO_2 emission rates induced by the higher taxation rates. The difference in the present value (worth) of the global GNP with and without a carbon tax (*i.e.*, Case A *versus* Case C) amounts to ~500 B$ lost from the world economy over the period of the computation, using a discount rate of 4 %/yr. The present value of the (consumption/users) carbon taxes collected at the Case-C rates, however, amounts to ~795 B$; this value does not simply "evaporate" from the world economy, but must "condense" in a sector(s) not modeled by the ERB model in its present form.

Although the shift in global concentration of economic capability caused by the imposition of a carbon tax is small, such a tax impacts more the developing countries[22]. The impact of this reduced energy usage coupled with the shift to reduce-carbon (oil and gas) or no-carbon (nuclear and solar) energy sources on the CO_2 emission rate is reflected in part by an increased use of NE, which impacts the degree to which the (world) uranium resource is depleted and (ultimately) the need to utilize advanced (breeder) fission systems if the NE option is to remain economical.

The benefits of reduced carbon emissions portended by imposition of a carbon tax (if an equitable and/or optimized collection strategy could be devised and implemented) must be balanced by the decreased GNP that results, as well as the increased potential for proliferation risk associated with greater global inventories of nuclear-explosive materials. The implications of a reduced GNP, which over the simulation period has a present value that is comparable or even less than the present value of the tax revenues collected, requires further study. The connection between risks associated with proliferation and risks associated with increasing GHG emissions and atmospheric accumulations can be examined though relative changes in an "operating curve" that (for example) relates PRI to the accumulation of carbon dioxide relative to present atmospheric inventories. Recognizing the risks of presenting oversimplified and possibly over-aggregated correlations, Fig. 17A gives the relationship between the global proliferation utility or proliferation-risk index and the cumulative CO_2 parameter GHP = $M_{CO_2}/M_{CO_2,1990}$ in going from the BAU Case A (no plutonium recycle, no carbon tax to Case C (no plutonium recycle, carbon tax), for a range of carbon-tax rates (Case C corresponds to a tax rate of 40 $/tonne/15yr). The imposition of a carbon tax results in a large decrease in this GHG risk metric parameter for a relatively small change in PRI.

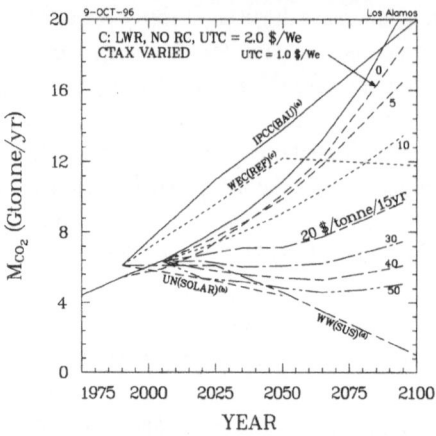

Figure 16. Time dependence of global CO_2 emissions as a function of carbon tax rate (proxy), also showing:· a) a range of other CO_2-emission scenarios; and b) the impact of reduced NE costs (Fig. 18).

(a) Refs. 51,52; (b) Refs. 52,53,58 and (c) Ref. 53

The PRI-GHP-tax relationship reported in Fig. 17A has been normalized in Fig. 17B to the Case-A conditions for each time; in this form the relative changes in PRI and GHP are better illustrated as a function of time and the rate at which the carbon tax is applied. The correlation depicted in Fig. 17 presents an opportunity to mislead if it is not recognized that the increased NE market share (and proliferation risk) results from fossil energy becoming more expensive *vis á vis* the carbon tax and not as a result of inherent economic attractiveness of NE; the reduction of GHP shown on Fig. 17 results from the carbon tax and reduced overall energy consumption and not because NE is assuming a larger market share. Lastly, irrespective of the driver, a means to translate these relative risk metrics into a relative "consequence space" is required before quantitative conclusions can be made on relative benefits *versus* cost related to nuclear (proliferation) and fossil (GHG impacts) fuels.

Nuclear Energy Costs

In the previous section, the market share for NE increased if the cost of fossil fuels is increased by the imposition of a carbon tax. The NE market share and composition (*e.g.*, LWR *versus* LMR) can also be shifted through changes in capital and fuel charges related directly to NE. In a sense, the impacts described on Fig. 16 cast the carbon tax into a role of proxy by which the costs of implementing CO_2-reducing technologies can be related to a given level of carbon dioxide mitigation without the actual imposition of a carbon-tax *per se*. This mode of assessment has been reported for a biomass-sustained hydrogen economy[12], and a similar analysis remains to be developed for NE (in the presence of and in competition with other sustainable energy supplies). Reference 22 describes the impact of NE economics in three areas: a) capital cost of LWRs; b) capital cost of advanced systems, like the LMR; and c) the cost of uranium *versus* accumulated consumption. Only results from the capital-cost variations are reported here.

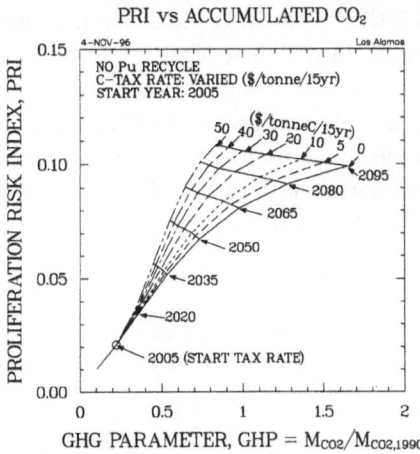

Figure 17A. Connectivity of risks metrics related to GHG emission and proliferation based on a Case A (BAU, no recycle, no carbon tax) *versus* Case C (BAU with a carbon tax rate of 40$/tonne/15yr) comparison; also shown is the impact of a range of carbon tax rates.

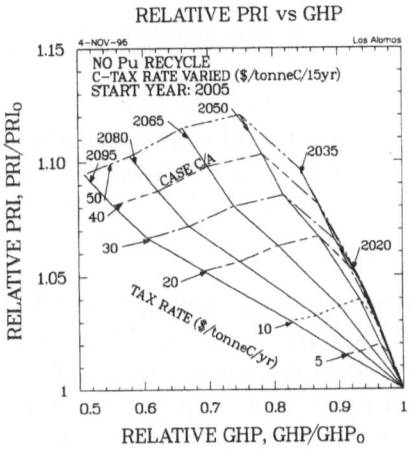

Figure 17B. Results given in Fig. 17A after normalizing at each time to the base-case (Case A) results.

LWR Capital Cost. The impact on NE market demand of increases or decreases in the unit total cost for LWRs ($UTC_{LWR} = 2.0$ $/We, conservative depletion model for uranium costs[28,59]) is illustrated in Fig. 18. All other NE unit costs remain as listed in Table II. A doubling of UTC_{LWR} decreases the global market share by a factor of ~4. The impact of this increased NE cost on GHG emission is relatively small; a similar finding emerges upon decreasing the cost of NE, as is shown on Fig. 18. In terms of total primary energy, NE contributes at most 7-9% of the global primary energy requirements (according to Ref. 45,

5.0% in 1990 and 5.9% in 2020, which differ somewhat from the ERB projections). The impact of carbon taxes on the NE market share (Case C, Table I) is also repeated on Fig. 18 for purposes of comparison. For NE to impact CO_2 emissions significantly (through the aforementioned carbon-tax proxy), non-electric applications (*e.g.*, process heat and/or electric power for hydrogen generation, for eventual use in the transportation end-use sector) must be considered. Generally, while high carbon taxes bode well for NEs market share, low NE electric-generation costs have little impact on global GHG emissions.

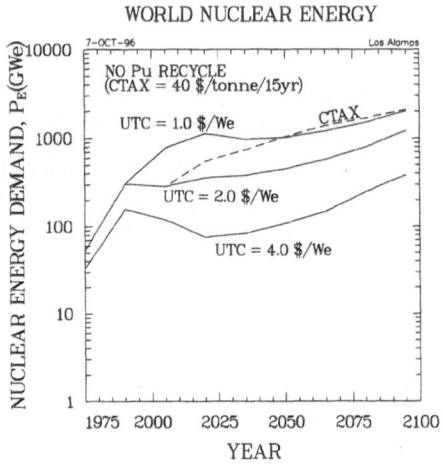

Figure 18. Impact of increased NE costs (Case B) or increased fossil-fuel prices (*vis á vis* a carbon tax, Case C) on the NE global market share (all cases are without plutonium recycle, Table I); also shown is the impact of decreased NE cost.

LMR Capital Cost. Heretofore, a large ratio of LMR to LWR unit total cost has been exogenously imposed ($UTC_{LMR}/UTC_{LWR} = 2.0$) to insure for computational purposes that the LMR would not be economically competitive with the LWRs, even for the conservative resource-depletion model used to estimate uranium prices[28,59]. It is emphasized that this ratio of LMR and LWR capital costs is an exogenous input and as such is independent of all input or computed (*e.g.*, uranium resource depletion) variables. Imposition of a strong carbon tax could allow the introduction of additional LWR-based NE to such an extent that the depletion of uranium fuel drives upward the price of uranium. The attendant increase in uranium prices, even with inventory-constrained plutonium recycle to LWRs, could allow the LMR to become competitive in the time frame of interest (*e.g.*, before 2100) if UTC_{LMR} was sufficiently low. Decreasing UTC_{LMR}/UTC_{LWR} to 1.3 while simultaneously imposing a strong carbon tax defines Case D. These conditions, along with the use of a pure depletion model for the uranium resource and cost, was sufficient for the LMR to become economically competitive by the year 2050. Figure 19 gives the fraction of global NE provided by LWRs as a function of time. A technology diffusion model[60] was used to limit the rate of LMR introduction to below that dictated by purely economic consideration; Fig.

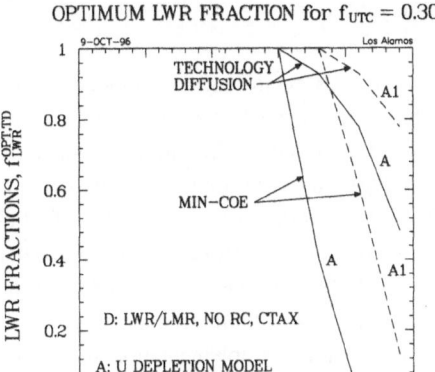

Figure 19. Market penetration of advanced (breeder) reactor under favorable Case D economic conditions (*e.g.*, conservative uranium-resource depletion model, competitive LMR costs relative to LWR base case, high carbon taxes).

19 gives both economics and technology-diffusion LMR introduction rates. The plutonium requirements for Case D are such that the global plutonium inventories stored in spent fuel become depleted by the year ~2065 for the introduction of high-inventory LMRs having unit breeding ratios. The impact of reducing the unit cost of uranium by a factor of two relative to the conservative resource-depletion model used in the base case is also shown on Fig. 19.

Generally, in the context of the ERB forced-economic-equilibrium model, the cheaper the fuel resource, the more of it will be used. Since the cost of uranium remains a relatively small part of the overall cost of NE, however, the impact of reduced uranium charges on the overall market share for NE remains small; a decrease in uranium prices by nearly a factor of eight over the base-case parameters results in less than a doubling of the NE capacity over the BAU Case A by the year 2100. The factor of ~8 reduction in uranium unit cost over that predicted by the resource-depletion model used herein gives a cost-*versus*-usage relationship that is similar to the one used in a recent study of the LMR[55]. Hence, with twice the base-case uranium-resource prices, the LMR under the Case-D conditions (strong carbon tax, $UTC_{LMR}/UTC_{LWR} = 1.3$) does not become economic until ≥ 2065; a reduction in uranium-cost *versus* resource relationship by a factor of ~8 stalls the LMR introduction to some time beyond the year 2100, based solely on economic (including resource) considerations.

It should be noted that the rudimentary NM balance presently used in the NE part of the ERB model fixes the LMR breeding ratio (BR = 1.0 in the cases reported here, Table II). Introduction of the high-specific-inventory LMRs at the economically driven, technology-diffusion-limited rates quickly depletes the accumulated (*e.g.*, all spent-fuel forms) plutonium inventories, and with the present level of NM "feedback", the global inventory of available plutonium goes to zero. This decrease combines with the reduced inventory of separated plutonium (the LMR is assumed to have a close-coupled, low-inventory reprocessing plant) actually diminishes the PRI, as most of the global plutonium is shift to (assumed) "safe and secure" in-reactor and in-processing inventories.

SUMMARY CONCLUSIONS

A global E[3] (energy/economics/environmental) model[8,12] has been adapted with an NE/NM model in an effort to understand better "top-level", long-term trade offs between civilian nuclear power, nuclear-weapons proliferation, fossil-fuel burning, and long-term global economic welfare. In addition to the implementation of a realistic NE/NM model, a proliferation-risk index (PRI) framework has been developed that shows promise for use in testing NM management strategies that minimize PRI or a related metric. The business-as-usual (BAU) point-of-departure base case[8] adopted for this study shows little NE growth in OECD countries, and, for the uranium resource model used, indicates little or no need for LMRs until well into the next century. Implementation of a strong carbon tax (rate) that increases the global share of NE, however, can under come circumstances increase NEs market share and deplete uranium resources to a point where the (economic) date for LMR introduction is advanced.

Using a BAU point-of-departure case, economic, resource, and proliferation-risk implications of plutonium recycle in LWRs, greenhouse-gas-mitigating carbon taxes, and a range of NE (capital and fuel) cost considerations have been examined. The parametric variations about the BAU base case are presented in the spirit of preliminary examples to provide vehicles for early critical review and comment. On the basis of these preliminary results it is suggested that, while GHG mitigation through some form of carbon taxation may increase the NE market share (as well as that for solar and other renewable energy sources), significant impacts of NE on the GHG problem will occur only if cost-competitive non-electric applications of NE are emphasized. Additionally, under conditions where in-reactor plutonium inventories can be considered "safe-and-secure", introduction of LMRs may lead to significant reductions in PRI, particularly if accompanied by a closely coupled integral fuel cycle. Under a similar assumption of "safe-and-secure" in-reactor conditions, the weights used to compute both the relative importance (*e.g.*, utility) of separated *versus* spent-fuel plutonium and the regional weights use to compute the global PRI (again, from the perspective of the a given country or region, in the case the U.S.) indicate little impact on the time evolution of PRI for the (global) inventory-constrained values of MOX-core fraction and related implementation times examined (Table II). Future work is aimed at refining the plutonium-inventory/flow model and extending the range of options/scenarios examined.

This study has identified important model limitations that require bounding, study, and development; these key areas of future work are listed below:

- Although a proliferation-risk metric has been evaluated in conjunction with a GHG-reduction metric, the crucial connection to "consequence space" remains to be made, wherein the actual costs of given levels of PRI and GHG accumulations can be compared on a common basis;
- The (subjective) MAU-based proliferation model provides a useful analytical structure, but bases for utility-function and weighting choices must be evaluated/elaborated; feedback mechanisms between NWA, ENV, and SANC parameters (Fig. 5) and proliferator utilities remain to be formulated;
- A focus for proliferation risk that is broader than the present (back-end) inventory-based civilian fuel cycle needs to be developed; comparisons of proliferation risks associated with other sources of NW-usable material must be made (*e.g.*, HEU, Fig. 5), as well as the other four proliferation drivers listed in Fig. 4;
- The use of the carbon-tax proxy as a means to assess the cost *versus* benefit of sustainable, GHG-reducing energy technologies, as well as technology-diffusion

limiting introduction rates of these technologies require additional model development and benchmarking;

- Metrics that relate energy security to national security need to be developed, along with the introduction of rudimentary measures of political/economic stability, and related risks (*e.g.*, NW accessibility as well as long-term capital availability; impacts of regional environmental, and resource limits on growth in human welfare, *etc.*)
- Improved geographical resolution, at least insofar as access to and assessment of each of the three fundamental choices available to address E^3 issues is concerned (*e.g.*, doing without; doing it better; or doing it differently)[61]; develop and implement economic elements that differentiate elements between decentralized and centralized economies, particularily as these elements reflect in the geo-political changes that have occurred since the ERB model first appeared[8].

The above list will provide guidance for future work on the analytic capabilities being developed to support the Los Alamos Nuclear Vision Project.

NOMENCLATURE

ACC	accumulated plutonium (as spent fuel in all forms)
AFR	Africa
ANS	American Nuclear Society
ASI	Advanced Science Institute (NATO series)
APU	accumulated plutonium (as spent fuel in all forms)
ATW	accelerator transmutation of (nuclear) waste
BAU	business as usual
BE	backend of fuel cycle
BN	actinide/fission-product burner
BR	breeder LMR
BU(GWd/tonne)	fuel burnup
BU25	^{235}U burnup
CE	uranium conversion/enrichment
CISA	Center for International Security Analysis
COE(mill/kWeh)	cost of electricity
CRT	criticality hazard
CSF	cooling for spent fuel
CST	cost
CST_{MAR}	marginal cost
CST_{TOT}	total cost
CV	$U_3O_8 \rightarrow UF_6$ conversion
D	detection (of proliferation action)
D&D	decommission and decontamination
DT(yr)	development time for NW capability
E^3	economics/energy/environmental
EEU	eastern Europe
ENV	political environment
EPA	Environmental Protection Agency
EPRI	Electric Power Research Institute
ER	enrichment

ERB	Edmonds/Reilly/Barns (model)[8]
e	political environment index
FC_k	fuel cycle of kth kind
FCR(1/yr)	fixed-charge rate
FE	frontend of fuel cycle
FF	fuel fabrication
FP	fission products
FSU	former Soviet Union
FTW	fusion transmutation of (nuclear) waste
f_{LWR}	fraction nuclear power provided by LWRs
f_{MOX}	MOX core (volume) fraction
f_{UTC}	UTC cost factor for LMR, $(UTC_{LMR} - UTC_{LWR})/UTC_{LWR}$
GDP	gross domestic product
GHG	greenhouse gases
GHP	greenhouse parameter, $M_{CO_2} / M_{CO_2,1990}$
GNP	gross national product
GR	global region
GRI	Gas Research Institute
HEU	high-enriched uranium
HM	heavy metal (uranium)
IEA	International Energy Agency, Institute of Energy Analysis (ORAU)
IIASA	International Institute for Advanced Systems Analysis
ITD_{MP}	inherent technological difficulty related to materials processing
ITD_{NW}	inherent technological difficulty related to nuclear weapons fabrication
i	material stream index
j	attribute, criteria index
k	fuel cycle index
LEU	low-enriched uranium
LMR	liquid-metal reactor
LP	linear program
LTS	large technical system
LWR	light-water reactor
l	country/region index
$M_i(kg)$	material inventory in stream i
MA	minor actinides
MAU	multi-attribute utility (analysis)
M_{CO_2} (Gtonne)	cumulative CO_2 emissions since beginning of computation (1975)
$M_{CO_2,1990}$ (Gtonne)	atmospheric CO_2 inventory in 1990 (742 Gtonne[56])
$\dot{M}_{CO_2}$ (Gtonne/yr)	global CO_2 emission rate
MIT	Massachusetts Institute of Technology
MM	mining and milling
MOX	mixed uranium/plutonium oxide fuel
MP	material processing
MS	monitored storage
m	time index
NE	nuclear energy
NM	nuclear materials
NU	natural uranium
NW	nuclear weapon

NWA	nuclear weapon aspiration level
n	number of assessment criterion, or NWA index
O&M	operating and maintenance (costs)
ORAU	Oak Ridge Associated Universities
OT	once-through LWR
P	production function or power
PNL	Battelle Pacific Northwest Laboratory
PRI_i	proliferation risk index for material stream i
PU	plutonium
p_f	availability factor
Q_E	engineering gain or Q-value
R, Rx	reactor
RAD	radiation hazard
RAF_j	risk-aversion factor for criterion j
RC	recycle (plutonium)
RCP	plutonium recycle LWR
RCU	uranium recycle LWR
REPRO	reprocessing
RP	risk potential or reprocessing
RPU	reactor plutonium inventory
RU	recycled uranium
r(1/yr)	discount rate
SANC	sanctions
SCI	science
SEI	Stockholm Environmental Institute
SEP	separated plutonium (reprocessing and/or fuel fabrication)
SF	spent fuel
SF_{NEW}	"fresh" spent fuel
SF_{STR}	spent fuel in short-term (~15-20 years) storage
SF_{RPC}	spent fuel in closed (sealed) repository
SF_{RPO}	spent fuel in open (operating) repository
SFT	total spent fuel
SI(kg/MWt)	specific inventory
SNM	special nuclear material
SOI	status of information
SPU	separated plutonium in processing and storage, excluding exo-reactor MOX
STR	storage
SW	separative work
TR	transport
$T_{MOX}(yr)$	half time for MOX introduction
TEC	technology
TH	thermal
t(yr)	time
UC_j($/unit)	unit cost of jth item
UNEP	United Nations Environmental Project
UTC($/We)	unit total cost
u	(general) utility function

	weighted averaged utility function for material stream i
$<u>_j(x_j)$	weighted averaged utility function for criterion j
$u_i(y_i)$	separable utility function for material stream i
$u_j(x_j)$	separable utility function for criterion j
$u_{ij}(y_i,x_j)$	component utility function
WP(yr)	warning period (as fraction of project remaining)
WPU	weapons plutonium
$w, w_{ij,i}$	utility function weights, normalized
x_j	mass fraction of jth component
$\lambda_j(1/yr)$	time constant for j^{th} process
η_{TH}	thermal conversion efficiency
η_p	plant overall efficiency, $\eta_{TH}(1 - 1/Q_E)$
$\tau_j(yr)$	characteristic (hold-up) time for j^{th} process
v	uranium costing exponent

REFERENCES

1.   W. K. Panofsky, Tension between nuclear proliferation danger, nuclear energy, and the disassembly of nuclear warheads, VIII Intern. Amaldi Conf., Piecenza Italy (October 5-7, 1995).

2.   W. G. Sutcliffe, Selected papers from Global'95 concerning plutonium, Lawrence Livermore National Laboratory document UCRL-ID-124105 (June 14, 1996).

3.   Management and disposition of excess weapons plutonium: reactor-related options, Committee on International Security and Arms Control, National Academy of Sciences, National Academy Press (1995).

4.   F. Berkout, The rationale and economics of reprocessing, p. 35, (Ref. 2).

5.   K. E. H. Wooley and A. D. Elsden, Why recycle?, p. 19, *ibid.*

6.   E. D. Arthur and R. L. Wagner, The Los Alamos National Laboratory nuclear vision project, Proc. of the Uranium Institute 21st Annual Symposium, London UK (September 4-6, 1996) (also, this conference).

7.   T. R. La Porte, Social Responses to Large Technical Systems: Control or Anticipation, NATO ASI Series, Vol. 58, Kluwer Academic Publishers, Dordecht, Germany (1989).

8.   J. Edmonds and J. M. Reilly, Global Energy: Assessing the Future, Oxford University Press, New York (1985).

9.   J. Edmonds and J. M. Reilly, Global energy production and use to the year 2050, Energy, 8(6), 419 (1983).

10. J. A. Edmonds, J. M. Reilly, R. H. Gardner, and A. Brenkert, Uncertainty in future global energy use and fossil fuel CO_2 emissions 1975-2075, US Department of Energy Report DOE/NBB-0081 (December 1986).

11. J. Edmonds and J. M. Reilly, A long-term global energy economic model of carbon dioxide release from fossil fuel use, Energy Economics, 74 (April 1983).

12. J. A. Edmonds, M. Wise, and C. N. MacCracken, Advanced energy technology and climate change: an analysis using the global change assessment model (GCAM), Battelle Pacific Northwest Laboratory report PNL-9798 (May 1994).

13. J. A. Edmonds, M. Wise, and D. W. Barns, The cost and effectiveness of energy agreements to alter trajectories of atmospheric carbon dioxide emissions, Energy Policy, 23(4/5), 309 (1995).

14. A. S. Manne and R. G. Richels, Buying Greenhouse Insurance: The Economic Costs of Carbon Dioxide Emission Limits, The MIT Press, Cambridge MA (1992).

15. LEAP long-range energy alternative planning system, User Guide for Version 94.0, Stockholm Environment Institute - Tellus Institute report, Boston MA (October 1993).

16. M. Lazarus, C. Heaps, and D. Hill, The SEI/UNEP fuel chain project: methods, issues and case studies in developing countries, Stockholm Environment Institute - Tellus Institute report, Boston MA (May 1995).

17. W. D. Nordhaus, Managing the Global Commons: the Economics of Climate Change, MIT Press, Cambridge Massachusetts (1993).

18. W. Haefele and L. K. Kirchmayer (eds.), Modeling of Large-Scale Energy Systems, Proc. IIASA/IFAC Symp. on Modeling Large-Scale Energy Systems, (February 25-29, 1980), Pergamon Press, New York (1981).

19. N. Nakicenovic (Study Director), Global energy perspectives to 2050 and beyond, World Energy Council report prepared by IIASA (August 1995).

20. R. A. Krakowski, Global energy modeling in support of understanding long-term nuclear (materials) futures, Los Alamos National Laboratory document LA-UR-96-1931 (June 5, 1996).

21. R. A. Krakowski, A multi-attribute utility approach to generating proliferation-risk metrics, Los Alamos National Laboratory document LA-UR-96-3620 (October 11, 1996).

22. R. A. Krakowski, Global economic/energy/environmental (E^3) modeling in support of the Los Alamos National Laboratory nuclear visions project, Los Alamos National Laboratory document LA-UR (to be published, 1997).

23. W. D. Nordhaus, The Efficient Use of Energy Resources, Yale University Press, New Haven, Connecticut (1979).

24. W. D. Nordhaus and G. Yohe, Future carbon dioxide emissions for fossil fuels, Changing Climate, pp. 87-153, National Academy Press, New York (1983)

25. M. A. Wise, private communication, Battelle Pacific Northwest Laboratory, Washington D.C. (1995).

26. J. F. Clarke, The cost and benefit of energy technology in the global context, Proc. Conf. Technology Responses to Global Environmental Challenges: Energy Collaboration for the 21st Century, p. 521, Kyoto Japan (November 6-8, 1991).

27. W. Leontief, Input-Output Economics, Oxford University Press, Oxford (1966).

28. P. Silvennoinen, Nuclear Fuel Cycle Optimization: Methods and Modelling Techniques, Pergamon Press, Oxford (1982).

29. Nuclear Proliferation and Safeguards, U.S. Congress Office of Technology Assessment, Praeger Publishers, New York (1977).

30. B. G. Chow and K. A. Solomon, Limiting the Spread of Weapon-Usable Fissile Materials, RAND National Defense Research Institute, Santa Monica, CA (1993).

31. M. M. May, "Nuclear weapons supply and demand," Amer. Sci., 82, 526 (1994).

32. T. B. Cochran, "Proliferation and the nuclear disarmament process," Energy Policy, 23(3), 195 (1995).

33. M. M. May and R. E. Avedon, The future role of plutonium, Summary of a Workshop held at Stanford University, Center for International Security and Arms Control (March 29-30, 1994).

34. G. T. Seaborg (Chm.), Protection and management of plutonium, American Nuclear Society Special Panel report (August 1995).

35. F. Berkhous and H. Feiverson, Securing nuclear materials in a changing world, Ann. Rev. Energy Environ. 18, 631 (1993).

36. G. T. Gardner, Nuclear Proliferation: A Primer, Lynn Rienner Publishers London (1994).

37. J. Pilat and S. Maaranen, The virtual proliferation of nuclear weapons: is it real? does it matter?, personal communication, Los Alamos National Laboratory document CISA-96-009 (January 26, 1996).

38. P. Silvennoinen and J. Vira, An approach to quantitative assessment of relative proliferation risks from nuclear fuel cycle, J. Oper. Res., 32, 457 (1981).

39. P. Silvennoinen and J. Vira, Quantifying relative proliferation risks from nuclear fuel cycles, Prog. Nuclear Energy, 17(3), 231 (1986).

40 I. A. Papazaglou, E. P. Gyftopoulos, M. M. Miller, N. C. Rasmussen, and H. A. Raiffa, A methodology for the assessment of the proliferation resistance of nuclear power systems, Massachusetts Institute of Technology report MIT-EL 78-021/022 (September 1978).

41. C. D, Heising, I. Saragossi, and P Sharafi, A comparative assessment of the economics and proliferation resistance of advanced nuclear fuel cycles, Energy, 5, 1131 (1980).

42. T. L. Saaty, A scaling method for priorities in hierarchical structures, J. Math. Psychol. 15, 234 (1977).

43. R. A. Krakowski, Global energy projections: preliminary considerations of population-GNP feedback, Los Alamos National Laboratory document LA-UR-95-4316(Rev.) (January 22, 1996).

44. J. P. Holdren and R. K. Pachauri, Chapter 4: Energy, J. C. I. Dooge, G. T. Goodman, and J. W. M. la Riviere (eds.), An Agenda of Science and Technology into the 21st Century (1991).

45. World Bank, World Development Report 1990, New York (1990).

46. WEC Commission, Energy for Tomorrow's World: the Realities, the Real Options, and the Agenda for Achievement, St. Martin's Press, Kogan Page Ltd., London (1993).

47. D. Anderson, Energy and the environment, Special Briefing Paper No. 1, Edinburgh, Scotland, The Wealth of Nations Foundations (1991).

48. C. Starr, M.F. Searl, and S. Alpert, Energy sources: A realistic outlook, Science, 256, 981 (May 15, 1992).

49. Japan Research Institute of Energy Economics, W. F. Martin, Trilateral commission global energy security study discussion paper, Washington Policy and Analysis, Inc. (September 10, 1995).

50. W. F. Martin, R. Imai, and H. Steeg, Maintaining Energy Security in a Global Context, Report to the Trilateral Commission, Triangle Paper 48, The Trilateral Commission, New York (September 1996).

51. W. Pepper, et al., Emission scenarios for the IPCC: an update prepared for the IPCC working group I (May 1992).

52. N. Lanssen and C. Flavin, Sustainable energy for tomorrow's world: the case for an optimistic view of the future, Energy Policy, 24(9), 769 (1996).

53. WEC Commission, Renewable energy resources: opportunities and constraints 1990-2020, London (1993).

54. P. Beck, Prospects and Strategies for Nuclear Power: Global Boon or Dangerous Diversion?, Earthscan Publications, Ltd. (1994).

55. W. Burch, E. Rodwell, I. Taylor, and M. Thompson, A review of the economic potential of plutonium in spent fuel, Electric Power Research Institute report TR-106072 (February 1996).

56. D. J. Wuebbles and J. Edmonds, Primer on Greenhouse Gases, Lewis Publishers, Inc., Chelsea, Michigan (1991).

57. A. Dean, Costs of cutting CO_2 emissions: evidence from `top-down' models, Proc. OECD/IEA Conf. on the Economics of Climate Change, pp. 25-42, OECD, Paris (1994).

58. T. J. Johansson, *et al.,* Renewable Energy: Sources for Fuels and Electricity, Island Press, Washington, DC (1996).

59. R. Lehtinen, P. Silvennoinen, and J. Vira, Long-term supply of uranium with an optimized exploration effort, Proc. ANS Topical Meeting on the Technical Bases for Nuclear Fuel Cycle Policy, Newport, RI (September 1981).

60. J. C. Fisher and R. H. Pry, A simple substitution model of technology change," Technological Forecasting and Social Change, 3, 75, (1971).

61. V. Smil, China's Environmental Crisis: An Inquiry into the Limits of National Development, M.E. Sharpe, Inc., N.Y. (1993).

U. S. POLICY ON SPENT FUEL REPROCESSING: WHERE DID IT COME FROM?

A. David Rossin

Center Affiliated Scholar
Center for International Security and Arms Control
Stanford University, Palo Alto, California 94305

THE POLICY

On April 7, 1977, President Jimmy Carter announced that the United States would defer indefinitely the reprocessing of spent nuclear reactor fuel. He stated that after extensive examination of the issues, he had reached the conclusion that this action was necessary to reduce the serious threat of nuclear weapons proliferation, and that by setting this example, the U. S. would encourage other nations to follow its lead.

Some other nations went ahead with reprocessing and breeder development, but high costs and loss of political support have delayed plans and limited investment in many nuclear projects. The U. S. never regained its technological lead in nuclear energy development, its own nuclear power program had already gone from orders to cancellations, and the dream of future energy security from breeder reactors faded away.

His Executive Order also announced that the U. S. would sponsor an international examination of alternative fuel cycles, seeking to identify approaches which would allow nuclear power to continue without adding to the risk of nuclear proliferation. But no new magic answer was found.

A BRIEF HISTORY

This story could begin with the Manhattan Project, or with President Eisenhower's "Atoms for Peace" speech in 1953, or the 1974 explosion of a nuclear device by India, or with the demise of the Joint Committee on Atomic Energy in 1976 or, as I have chosen to do in this presentation, to jump directly to the 1976 presidential campaign.

Technology for Global Economic and Environmental Survival and Prosperity
Edited by Kursunoglu *et al.*, Plenum Press, New York, 1997

Seemingly in response to Carter's campaign speeches, the Ford administration took up the issue, and prepared a closely held inter-agency memorandum (the "Fri Report"). Seven weeks before the election, Robert Fri, National Security Advisor Brent Scowcroft and his aide, David Elliott, briefed the President on the issues. Fri's report presented the White House with three alternatives. Ford listened, and concluded that controlling proliferation had to be the highest priority, and that a temporary hold to work out the necessary safeguards would be a good resolution. But he knew that the issue was complex, and that what we did would have international ramifications.

The preparation of Ford's policy statement involved seemingly endless numbers of drafts and comments, editing sessions that went long into the night, and high-level arguments about international security, energy and political factors. Ford's statement was finally presented in a campaign speech at Portsmouth, Ohio, just five days before the 1976 election. He featured his position that control of nuclear proliferation had to take precedence over commercial and even national economic interests. He called for a delay of up to three years in starting the Barnwell reprocessing plant. However, all parties knew that a programmatic environmental statement was going into public hearings that could easily take two years, and that the "back end" of the Barnwell plant that was required to prepare the waste for disposal was the government's part, and it hadn't yet been designed, let alone built.

Some argue that it was Ford who actually stopped reprocessing, not Carter. Industry experts who analyzed the Ford statement concluded that it had been carefully constructed as a compromise between the "go ahead" and "stop" alternatives of the Fri Report. It was carefully worded so as to take a tough stance against proliferation and reprocessing, but with the flexibility (if Ford were re-elected) to proceed with Barnwell, once licensing and safeguards issues already under review were successfully resolved.

Carter's statement was totally different in tone. It said that reprocessing should not proceed, not only in the U. S. but worldwide, because it was not essential for foreseeable economic or uranium resource purposes. Therefore, his advisors reasoned, since it added to proliferation risks, "it just didn't make any sense to allow reprocessing to proceed." The U. S. position was stated as firm and final, and it expressly included a plan to explain it to the other nuclear nations in order to convince them to adopt it as well.

In a Sept. 25, 1976 speech in San Diego, Carter had promised that he would stop Barnwell until it was "needed" and safe, and only ever allow it to operate if it were on a multi-national basis. Each criterion was a potential show-stopper.

Specifically, President Carter stated that the United States would defer indefinitely its licensing efforts and all further commercial work on nuclear fuel reprocessing. The Executive Order also eliminated support for further work to support the Barnwell reprocessing plant or for the construction of the Clinch River Breeder Reactor, and reformulated ERDA's research priorities away from nuclear energy. These were budget actions that the administration could take on its own, and they effectively eliminated any possibility of operating the Barnwell plant.

In contrast to the lengthy formal proceedings on the future of the breeder reactor and the extensive public proceedings on every nuclear power action, (and particularly on the use of mixed-oxide fuels to be obtained from reprocessing, called GESMO), no environmental impact statement on the Carter policy was prepared, let alone was it debated or subjected to public hearings. The secrecy which surrounded the policy's formulation and the speed with which such a major policy decision was reached (only 2-1/2 months after his inauguration) hardly raised a ripple in the national media. But it was a shock to the entire energy industry, let alone to Barnwell's owners.

In President Carter's statement, he said that his policy was designed to set an example to the rest of the nations of the world to turn away from technologies that could lead to further nuclear weapons proliferation. Some nations (particularly France, the United Kingdom and Japan) continued their programs to establish commercial reprocessing. However, U. S. pressure did affect a number of specific decisions by various governments, and also encouraged opposition parties and activist groups in these and other nations where nuclear energy became a political issue.

By April 1977, several committees in the Congress were already moving on the issue, and the scope of some of the bills soon exceeded even that of the administration. Within a year, all further work on GESMO had been canceled, the Nuclear Non-Proliferation Act was before Congress, and the International Nuclear Fuel Cycle Evaluation (INFCE) in which some fifty nations participated had become a platform for extensive analysis and debate. By the end of the Carter administration in 1981, the NNPA was law, the nine-volume report of INFCE could be found gathering dust on countless bookshelves, and U. S. influence in international nuclear affairs had receded significantly.

In early 1982, President Reagan effectively rescinded the Carter policy, allowed programmatic approvals for reprocessing of U. S. origin fuel by the Euratom nations and Japan, and even said that reprocessing could again be considered in the U. S. But by that time, all the remaining industrial momentum in the United States was gone, and no serious effort to revive commercial reprocessing has emerged since then. Commercial reprocessing is a reality in Europe and Japan, but its progress has certainly not been aided by any U. S. administration. In September 1993, President Clinton stated that it was the policy of the U. S. not to reprocess, but that we would not interfere in the programs of other nations.

WHY RE-EXAMINE THIS HISTORY?

The issues that shaped the Carter policy decision are with us now and will continue to arise in the years to come. The nuclear weapons nations have bombs and components made of U-235 and plutonium that they no longer need for weapons. The U. S. government is in trying to establish a program in agreement with Russia for effective disposition of tons of excess weapons material so that it will not again become a threat for weapons use. One obvious option is to burn it in power reactors as mixed-oxide fuel.

Although most of the debate has been limited to the political science, international policy and technical communities, broader insights may be useful as these same issues take political shape in the coming years. At least, if major policy decisions need to be revisited, it might be useful to have, out in the open, a broader understanding of the facts and arguments which impacted these policy decisions.

A RE-EXAMINATION OF THE REASONING

During the 1976 campaign, Jimmy Carter had made two major speeches in which the issue of nuclear proliferation was raised to a high priority level. The source of these statements is important. For a year, a group of distinguished scholars had been working on a report with support from the Ford Foundation titled "Nuclear Power Issues and Choices." This book, when it was published in March, 1977, became more popularly known as "The Ford/Mitre Report" or just "the black book."

The Ford/Mitre book was provided to President Carter just in time to serve as the basis for his nuclear policy. However, drafts of its conclusions and recommendations had been furnished to people on the Carter transition team, and some members of the panel were appointed to key positions in the new administration. Two Ford/Mitre panel members had worked on the critical sections of Carter's campaign speeches, and they were soon tapped for important positions in the new administration.

While the panel's work was scholarly, its findings were not tested by, nor subjected to, debate by persons who had deep, working knowledge about the issues. In fact, that step was specifically avoided by the group that produced the book. It was their position that they were knowledgeable but also open-minded persons, and none of them were either industry partisans or anti-nuclear activists. They felt confident that they could listen to presentations by various interests and read the pertinent literature, and then when they arrived at their conclusions, there would be no further need to expose them to outside review before providing them to the President as the most scholarly and unbiased analysis of the entire subject.

The actual framing of the Carter policy took place in secret. When it was announced, it really was a complete surprise to the international nuclear energy industry. Certainly, some clues and signals had been seen, and Ford's statement had served warning, but the sweeping scope and rigidity of the Carter statement, and its effect of changing three decades of planning, Federal support and national policy, did come as a shock.

In fact, at the very same time it was announced, an advisory committee on the breeder reactor and nuclear energy, called for by the new President, was busy drafting their findings and recommendations to him. The committee was split 13 - 4. The President never even saw the majority report that recommended going ahead with reprocessing and the breeder reactor.

President Carter had approved the preparation of a highly classified Presidential Review Memo (PRM-15). It made these points:

a.   Prevent the spread of nuclear explosive-- or near explosive -- capabilities to countries which do not now possess them.
b.   Seek a pause among all nations in sensitive nuclear developments and provide incentives for nations to participate in an international search for alternatives.
c.   Study measures for international spent fuel storage and disposal, but not plutonium storage, which the U. S. shall discourage.

d.   Prepare a program for promoting the development of non-nuclear energy alternatives and for assisting other nations with non-nuclear means to meet them.

The four-page Presidential Decision memo (PD-8) was prepared by the staff of the National Security Council. Dated March 24, 1977, PD-8 outlined options for the President: Deputy Undersecretary of State Joseph Nye, who had served on the Ford/Mitre panel, approved three options for the President to consider, which were supported by different agencies:

1. Allow reprocessing in U. S. to start when safety and safeguards are assured. Explore multinational arrangements. [Nobody]
2. Stop reprocessing in the U. S. and approach other nations to follow our lead. [ERDA, Defense, State]
3. Stop all reprocessing everywhere and renegotiate all agreements to give the U. S. a veto on reprocessing. [ACDA, NSC]

He felt that he had found a consensus among the departments for the second option. However, Carter selected the third. (The policy as adopted allowed for the President to waive the total renegotiation requirement if he determined that it was necessary for U. S. national interests.)

FEAR OF NUCLEAR WEAPONS DROVE POLICY

The key issue driving the policy was the threat of proliferation by diversion of plutonium from the civilian fuel cycle. The U. S. had already discouraged South Korea and Taiwan from embarking on reprocessing programs. President Ford had initiated diplomatic efforts to dissuade France and Germany from proceeding with deals with Pakistan and Brazil respectively which included reprocessing or enrichment facilities.

Regardless of statements by these nations that their interest was only for civilian nuclear power, the U. S. had serious doubts. Also, U. S. policymakers were dubious that any safeguards regime could prevent diversion of material for weapons purposes.

Even before the 1970's it was known by experts that it was possible to make an explosive nuclear device from a wide range of mixtures of plutonium isotopes. In short, reactor-grade plutonium is not the material of choice for weapons, but it is not proper to argue that certain governments which may want nuclear weapons desperately, will not (or have not!) tried almost any possible route to get them. Efficiency, reliability and cost have not always been decisive factors. They would be, however, for an advanced industrial society if its objective were to build an arsenal of actual weapons to defend their nation or to attack another.

Information which showed that a nuclear explosion was achieved at the Nevada Test Site in 1962 using "reactor-grade" plutonium impressed those who knew of it. However, the important engineering and reliability reasons that had led all nations away from this highly irradiated material were discounted by Carter's advisors. The theoretical possibility was enough to drive their thinking to the conclusion that all stocks of separated plutonium had to be eliminated.

Nuclear experts generally took the position that a number of more attractive routes to nuclear weapons existed, and that these were the areas that called for real concern and action. But other voices entered the top-level debate. In addition to the Ford/Mitre panel, Albert Wohlstetter's writings focused on diversion from stockpiles of separated reactor-grade plutonium. The scenarios involved advance preparation of all components and expertise, followed by diversion and fabrication in such a short time that there would not be timely warning for diplomatic actions to prevent the emergence of a new nuclear weapons nation. It is obvious that reports of the Wohlstetter group contributed to the reasoning of people working on both the Ford/Mitre report and the Carter policy.

The Ford/Mitre panel invited various experts from the weapons laboratories, activist groups and the nuclear industry to present views and answer questions, but there is no evidence that they ever subjected their conclusions to outside review or debate by any of these persons. They were satisfied that they had found an appealing solution, a wise, and almost obvious compromise that should solve the major dilemma without creating any major dissatisfaction with the affected constituencies. They discounted warnings in testimony, presentations, and written reports that their proposed compromise could have serious negative impacts.

Interviews with Spurgeon Keeny, the Chairman of the Ford/Mitre panel, other members of the distinguished panel, and three of the main staff people who worked on the report, help explain its findings. The chapters were written by panel members themselves, a truly remarkable accomplishment for a high-level panel. The conclusions were written by Keeny and reviewed at the final meetings. All of the panel members felt that they could live with the conclusions, though some had different views on particular aspects of them. Several of the chapters, including the ones on economics and proliferation were extensively edited, and some practically rewritten, by Keeny and the staff. Keeny was very clear on the objective of the study: it was to give policy input to the newly elected president on nuclear energy and proliferation issues.

As Tom Neff, the staff director, who was a graduate student at MIT, explained, "I guess if you are a technocrat, you believe you can do it [reprocessing and waste disposal], but if you don't know entirely what the problems are, what would we rather do? Store spent fuel till we figure it all out, or do you want to start reprocessing and then try to figure out how to get rid of the products later?" He felt that spent fuel was a good waste form, so the closed fuel cycle should not have been a necessity in the first place. Neff also explained that the panel members themselves were already in high places and wanted to be definitive and to hit hard on issues. "I think, actually, in a policy environment, if you waffle around too much, nobody pays any attention to you."

Larry Ruff, the Ford Foundation project officer, also served as staff, and he actually rewrote the economics chapter. He recalls that in the final discussions, decision analysis played an important role. It was found by extensive economic modeling studies that though energy costs would represent a large absolute number over time, it was a small fraction of the total economy, and the panel determined that it should not be a crucial factor in ultimate decisions.

Most telling however, was the conclusion of the modeling studies that the cost of delay of 10, 20, or even 30 years had an almost insignificant effect on the overall economy. Thus the recommendation to defer reprocessing and the breeder was easy, because it would carry no penalty discernible to the general public. (I found no similar decision analysis process on the alternative routes to nuclear weapons.)

As soon as the book was published, the entire panel was invited to an Oval Office meeting with President Carter. He had read the book the night before, and asked detailed questions. The conclusions were obvious:

* Reprocessing: On basis of arguments presented to Carter, it didn't make any sense.

* The Ford/Mitre panel thought they had found a simple and obvious compromise.

* The opinion was that nuclear power could go ahead successfully without reprocessing.

Even the fact that eliminating reprocessing would only affect one possible route to proliferation, and that others were easier to conceal, cheaper and more reliable, did not deter the Carter strategists from their final narrow policy choice. This rigid policy carried serious downside risks of greater long-term dependence on oil, as well as coal and gas, and the loss of America's technological lead and credibility with our friends and allies.

Missing from U. S. policy thinking has been differentiation between totally different types of threats: industrialized nations, nations outside the NPT or suspected of subverting it, and subnational or terrorist groups. The Carter policy was an attempt to solve all potential problems with a single comprehensive policy initiative. However, history suggests that the issue is more complex than the policy's proponents admitted.

The book itself was so new that there had been no time for public review or comment, or for any kind of feedback on it to the new administration. How could Carter be so aware of complexity of problems and be satisfied with simplistic solutions? How could his moral code justify breaking so many international commitments? The Administration featured talking with other nations, but the Ford/Mitre panel never listened to what they said.

Proliferation threats were not differentiated. Jimmy Carter wanted a comprehensive policy that solved all proliferation problems. The goal was a leakproof regime, which was unrealistic in view of all the alternative paths to nuclear weapons. Academics saw this issue as an opportunity to contribute to world peace and to be part of a major policy change.

Europeans believed that U. S. with its huge resources was trying to get an advantage and recover its loss in world leadership in nuclear energy. They believed this, because to them, the Administration's economic and proliferation arguments did not make any sense. One key question was the credibility of a U. S. policy that might not treat all nations alike. In both the Ford statement and that of Carter, the idea of discriminating between nations was stated to be unacceptable. Ironically, examples to the contrary are the norm!

ECONOMICS AND ARITHMETIC

Economic analyses played a decisive role in reaching the decision to stop reprocessing. A key position of the Fri Report and the Ford/Mitre panel was that reprocessing offered little economic benefit, and therefore, the reasoning went, if it carried any additional proliferation risks, it was logical to prevent it. Actually, most of the analyses did show a break-even or slight advantage to reprocessing relative to mined and enriched uranium fuel costs. However, fuel cost was only 1/4 of nuclear power generation cost, and generation cost was less than half of the consumer's electricity price. Furthermore, nuclear power was competing with oil and gas, both at much higher costs, and at that very time, the government was urging utilities to cut back on their use, so any economic savings appeared unimportant.

More important was the fact that utilities were seriously concerned about long-term factors, like nuclear waste disposal and ultimate utilization of plutonium in breeders, which made closing of the nuclear fuel cycle very important to their long-range planning. The arithmetic used in calculating relative fuel costs may well have been right, but of more serious consequence, these more important factors were not included in the decisive "economic" analyses.

CARTER THE PRESIDENT -- CARTER THE MAN

Historians and political scientists have observed that President Carter studied individual issues in great detail. Time after time he looked for comprehensive solutions. Once he had weighed the facts he had available to him (and he had reason to believe he had them all) he would make his decision. From that point on, he held it rigidly. Far from being indecisive as some observers claimed, he stubbornly held to his positions and did not tolerate further attempts by Cabinet secretaries or staff aides to get him to make any changes. In explaining decisions, he spoke of moral reasons and of doing the right thing despite potential political and economic penalties.

Jimmy Carter was truly concerned about proliferation. He made reducing the threat a personal commitment and an objective for his administration. He believed that steps in that direction were "the right things to do." Interestingly, though the general commitment to non-proliferation goals appeared in certain campaign speeches and in the list of "promises" he wished to fulfill, concerns about the nuclear fuel cycle, reprocessing, and even the Clinch River Breeder Reactor, rarely appear on any list of high-priority issues identified by those who have chronicled and studied the Carter presidency.

It is fair to ask if this kind of decision scenario could have occurred under a different president than Jimmy Carter. Arguably, no greater opposites could be found in their approach to decisionmaking than Jimmy Carter and Gerald Ford. Carter's analytic approach, his appetite for details, his distaste for business, his distrust of the bureaucracy, and his determination to do what he saw as right even if it would be the hard way, showed through in any number of his decisions. This decision came out the way it did in no small part due to Carter being the President. But could this happen with others? Of course! It has, and it will again!

WHERE WAS THE NUCLEAR ESTABLISHMENT?

Among the people who drafted the Ford and Carter policies there were a number who fully believed that nuclear power could proceed in a healthy way without reprocessing. They felt that nuclear power could even be more acceptable to the public if it could free itself of the image of proliferation danger. At the same time, the leading anti-nuclear activists fully recognized that without reprocessing, the nuclear power fuel cycle could not be closed, the breeder reactor would be stopped, and nuclear power's future would be limited if not ended. This fit their own personal and organizational goals.

A critical analysis of the positions taken by various groups in the political battles on this issue reveals that despite stated concerns about economics and nonproliferation, the actions called for would fit well with a serious political initiative to eliminate nuclear energy as a major source of electricity for the future. Leaders of the global nuclear industry recognized the single-mindedness and dedication of the activists and the seriousness of this threat, and felt pushed to the wall by it. They complained privately and publicly, but their arguments about relative risk, difficulty (rather than impossibility of proliferation from civilian reprocessing), adequacy of safeguards (though not perfection), and long-term needs for energy failed to win attention. There was little public interest and hardly any political support.

At the same time, those involved directly with commercial nuclear power had credibility problems of their own. Their concerns related to business enterprises at a time when the public mood was not supportive of business. Also, many nuclear scientists and researchers felt that they should stick to their scientific work and stay out of the policy and political arenas.

The newly formed ERDA was not regarded as a strong agency. It was lame duck, soon to be replaced by the new Department of Energy. The old AEC and the nuclear power industry itself, along with its few supporters in Congress (holdovers from the former Joint Committee on Atomic Energy) were seen as being locked into rigid beliefs and fearful of any change to their traditional positions. Therefore, it was felt that their views could conveniently be ignored without political penalty.

The policymaking groups under both presidents, while having access to all the expertise available in government, the military and the National Laboratories, were very selective about the inputs they used. ERDA's experienced nuclear people were effectively excluded from policymaking, both under Ford and Carter, because they were viewed as high-handed and inflexible, being supporters of the conventional wisdom of what was required for a viable nuclear future. (It is not surprising that when asked to give up the accomplishments and hopes of thirty years of development in order to reconsider issues that had been argued out and resolved long before, experienced experts were less than eager to do so.)

This was a polarized debate, rarely face-to-face, while underlying objectives were not visible. No simple compromise was even possible. The nuclear industry was perceived as rigid. But if same test were applied to the tactics of anti-nuclear activists, they would appear rigid in demands that could stop nuclear energy, and they had responsibilities to no one. At the same time, the energy industry was listening to Carter's arguments on the energy crisis. To them, it seemed that undermining nuclear energy didn't make any sense. In fact, the Carter policy statement was within weeks of his energy speech in which he identified the energy crisis as "The Moral Equivalent Of War" (the MEOW speech). He did not, however, say that the war was important enough to revitalize nuclear energy.

SINCE 1977

State Dept. professionals were appalled as concerns of other nations were swept aside and legislation was passed that locked the U. S. into its own new policy. President Carter viewed the INFCE study as the vehicle for buying time, and providing an opportunity for other governments to reconsider their views and, hopefully, accept the U. S. position. In fact, it bought (or really, it cost) three years, during which all the American commercial trust and interest were dissipated. It did not change the minds of the other nations. (It seemed as if the Carter policy people never entertained the idea that perhaps we might listen more carefully to what the others brought to the table.)

This issue brought certain political forces that were generally polarized against one another into an alliance that had little forceful opposition. (One astute observer characterized it as a "horseshoe issue.") By 1978, Congress had become even more negative on nuclear issues than the administration. It seemed as if any attack by an environmental or political activist energized the Congress, and there was little interest on Capitol Hill in the views of industry executives who actually had responsibilities to their customers for reliable electric power, or even of Carter administration officials who began to realize that they had lost control of the legislation.

Many changes have taken place in the world energy picture, and the role of nuclear power has declined from all of the projections that were being made at that time. Much more is understood about nuclear weapons. More mature views of their limited military role and of the threat of proliferation have emerged. True, the republic has survived, and energy issues are off the front page. Proliferation threats still make the newspaper on occasion, but the Cold War is over, and the fear of all-out nuclear war no longer dominates children's dreams.

The nuclear fuel cycle continually resurfaces as an issue for debate in the U. S. When it again becomes a major policy battle, it will be vital that the issues be clearly faced, and that myths and half-truths that affected decisions in 1976 and 1977 no longer be accepted without debate. The availability of reasoned, documented arguments could be of immense importance to any informed segment of the public, and could become a factor in the political decisions that are ultimately reached. At a minimum, we should be able to raise the quality of that future debate well above that of two decades ago.

CHAPTER V
NUCLEAR TECHNOLOGIES

BORON NEUTRON CAPTURE THERAPY OF MALIGNANT BRAIN TUMORS AT THE BROOKHAVEN MEDICAL RESEARCH REACTOR

D.D. Joel, J.A. Coderre, and A.D. Chanana

Medical Department
Brookhaven National Laboratory
Upton, NY 11973

INTRODUCTION

Boron neutron capture therapy (BNCT) is a bimodal form of radiation therapy for cancer. The first component of this treatment is the preferential localization of the stable isotope ^{10}B in tumor cells by targeting with boronated compounds. The tumor and surrounding tissue is then irradiated with a neutron beam resulting in thermal neutron/^{10}B reactions ($^{10}B(n,\alpha)^7Li$) resulting in the production of localized high LET radiation from alpha and ^7Li particles. These products of the neutron capture reaction are very damaging to cells, but of short range (each less than 10 μm) so that the majority of the ionizing energy released is microscopically confined to the vicinity of the boron-containing compound. In principal it should be possible with BNCT to selectively destroy small nests or even single cancer cells located within normal tissue. It follows that the major improvements in this form of radiation therapy are going to come largely from the development of boron compounds with greater tumor selectivity, although there will certainly be advances made in neutron beam quality as well as the possible development of alternative sources of neutron beams, particularly accelerator–based epithermal neutron beams.

Clinical trials of BNCT in the treatment of brain tumors were conducted at the Brookhaven Graphite Research Reactor between 1951 and 1959 (28 patients) and subsequently at the specially designed Brookhaven Medical Research Reactor (BMRR) between 1959 and 1961 (see review by Slatkin[1]). These early clinical trials were considered largely unsuccessful in that; there was no evidence of life extension; nonhealing ulceration of the skin developed in some patients; and in an effort to increase the radiation dose to deep seated tumors, four patients died as a result of the therapy.

It is generally believed that two major deficiencies led to the disappointing results of the first clinical trials. (1) The use of low-energy thermal neutron beams resulted in the peak neutron flux at the surface of the head, followed by an exponential decline so that the neutron flux at tumor depth was insufficient. (2) The compounds employed at that time did not provide the required selective accumulation of boron in the tumor. In this context, two important developments have occurred since 1961. First, through the use of appropriate filters and modulators the mean

Technology for Global Economic and Environmental Survival and Prosperity
Edited by Kursunoglu *et al.*, Plenum Press, New York, 1997

251

energy of the neutron beam at the BMRR was made slightly higher[2,3], concurrent with the suppression of the thermal neutron flux. This modified beam, referred to as an epithermal neutron beam, results in a lower incident thermal neutron flux at the surface of the head, with increased thermal neutron fluxes at depth in the brain.

A second major advance has been the development of boron compounds which yield greater concentrations of ^{10}B in the tumor as compared to those in surrounding normal brain tissues. Although several classes of boron delivery agents are in various stages of development (see review by Barth, et al.[4]), including boronated porphyrins, nucleosides, amino acids, polyamines, monoclonal and bispecific antibodies, liposomes, and growth factors, only two boron compounds currently are being used clinically, sodium borocaptate (BSH) and p-boronophenylalanine (BPA). BPA was originally synthesized in 1957 for potential use in BNCT of rapidly growing neoplasms[5]. It is an analog of the amino acid tyrosine, a precursor for melanin synthesis, and has been employed by the Japanese in the BNCT treatment of melanomas[6]. The use of BPA as a boron carrier for BNCT of malignant brain tumors was largely developed at Brookhaven using animal tumor models, particularly a transplantable malignant brain tumor in rats referred to as the 9L gliosarcoma[7-9].

BPA-Based BNCT Preclinical Investigation

BPA is a non-toxic, metabolic compound that is actively transported across the blood-brain-barrier (BBB) and therefore has the potential for accumulation in islets or streamers of tumor cells that are otherwise protected by the BBB. It is thought that these islets of tumor cells lying outside the main body of the tumor are the sites of the majority of recurrences following conventional radiation and/or chemotherapy.

Because of its insolubility in aqeous solutions, BPA was first given orally. Later methods were developed, first by the Japanese[10] and then modified by scientists at Idaho State University[11] and Brookhaven[9], to solubilize BPA by complexing it with fructose. Currently, the BPA-fructose complex (BPA-F) is administered systemically to both animals and humans.

When rats bearing intracerebral gliosarcomas are injected with BPA-F, tumor boron concentrations greater than 50 μg/g are readily achieved with tumor-to-blood and tumor-to-brain ratios of about 4 to 1. The biochemical mechanism(s) responsible for the selective uptake of BPA in cells of primary brain tumors is not fully understood but may, in part, be related to their high metabolic activity. BPA, is an animo acid analog, and is actively transported across the tumor cell membrane by the neutral amino acid transport system.

When the heads of tumor-bearing, BPA-F injected rats are exposed at the optimal post-injection time to neutron irradiation (BNCT), long-term survivals (cures) were obtained in the absence of demonstrable damage to normal brain, i.e., selective ablation of the malignant brain tumor[12]. The fraction of rats "cured" by BNCT (as high as 90%) was radiation-dose dependent, with a steep response seen between 22.5 Gy and 60 Gy-Eq (see Figure 1).

Clinical Investigations

After extensive studies in animals on the efficacy and toxicity of BPA-F, an FDA-sanctioned protocol to study the biodistribution of BPA-F in human patients scheduled to undergo surgical debulking of their malignant brain tumor was initiated in January, 1994. These studies were done in collaboration with the Beth Israel Medical Center in New York City. Patients were infused intravenously with BPA-F

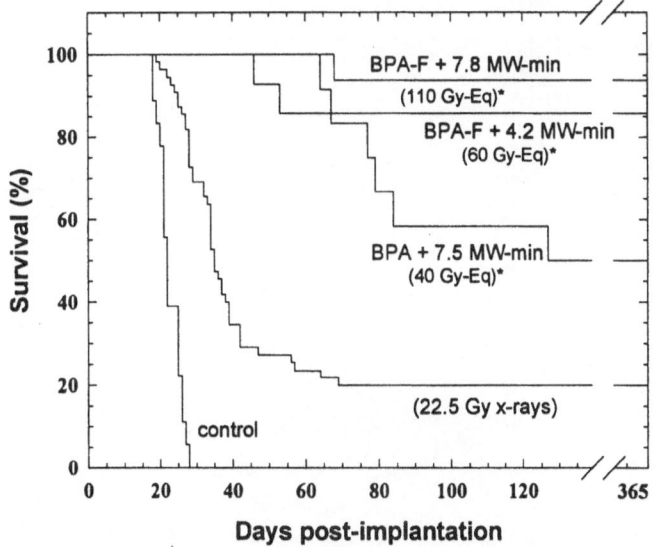

Figure 1. Survival of rats bearing intracerebral 9L gliosarcomas as a function of time after tumor implantation. The median survival of untreated controls (n = 18) was 22 days. All irradiations were performed on day 14 after tumor implantation. Rats treated with 22.5 Gy of 250 kVp X rays (n = 55) had a median survival of 35 days with 20% long-term survivors. Rats treated with 7.5 MW-min of reactor irradiation following oral administration of BPA (BPA + 7.5 MW-min, n = 12) showed 50% long-term survivors. Rats treated with PPA-F and either 4.2 (n = 14) or 7.8 (n = 14) MW-min of reactor irradiation showed 85% and 93% long-term survivors.

*Gy-Eq is equal to the physical absorbed dose (Gy) times an experimentally determined biological effectiveness factor for each dose component[13], including boron capture [^{10}B(n,α)^{7}Li], fast neutrons [^{1}H(n,n')p], nitrogen capture [^{14}N(n,p)^{14}C] and gamma rays.

over the 2-hour period just prior to the beginning of surgery. Samples of tumor removed by the neurosurgeon were analyzed for boron content and sections of these specimens were taken for histopathology.

In the first group of subjects studied, it was observed that boron concentration in tumor samples varied considerably among patients and even within multiple samples from individual patients. Histologic sections prepared from specimens analyzed for boron suggested there was a correlation between the degree of tumor cellularity and boron concentration[14].

This correlation was quantified in a total of 14 patients, receiving varying doses of BPA and the results suggested that the ^{10}B uptake in the active part of malignant brain tumors was quite consistent from patient to patient[15]. From these data we calculated that following the intravenous infusion of 250 mg BPA/kg (the dose currently used to treat patients), the ^{10}B concentration in the most cellular regions of malignant brain tumors in humans is about 50μ/g, with tumor-to-blood and tumor-to-brain concentration ratios of ~4 to 1. These ratios are similar to those found in rats.

The first patient with glioblastoma multiforme was treated with BNCT in September, 1994 i.e., 33 years after the termination of the early clinical trials (Figure 2). After a four-month period of observation to verify the safety of the procedure, particularly with regard to the potential for early-delayed neurological problems, a multipatient protocol was initiated in February 1995[16]. The objectives of the first protocols (a slightly modified protocol was started in 1995) were: (1) to determine a

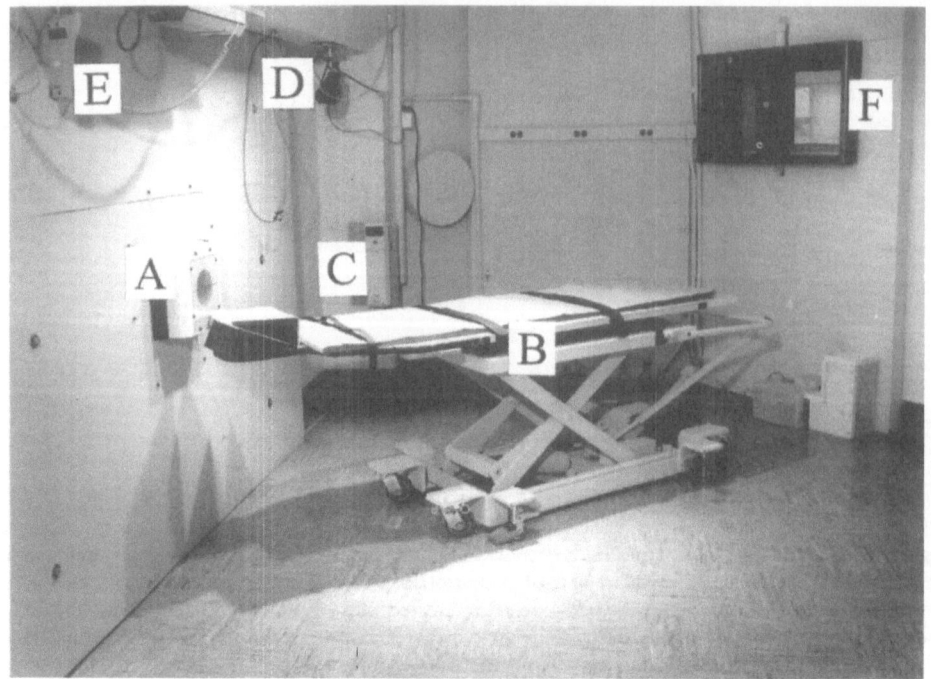

Figure 2. BNCT treatment room at the Brookhaven Medical Research Reactor. (A) Epithermal neutron beam port and collimator; (B) Patient gurney; (C) Laser beams for patient alignment; (D) Video monitoring; (E) Voice communication; (F) Observation window.

safe starting dose for BNCT; (2) to evaluate any adverse effects of BNCT and; (3) to evaluate the effectiveness of BNCT at a safe starting dose in patients with glioblastoma multiforme. These protocols were terminated in February, 1995, after 15 patients had been treated with BPA-based BNCT at the BMRR.

The major conclusions drawn from these 15 patients were: (1) BNCT, as administered to these patients, was safe. There were no adverse effects associated with the infusion of BPA-fructose at a dose of 250 mg BPA/kg body weight. There was no damage to the scalp other than focal alopecia and no damage to normal brain or other critical organs was observed; (2) tumor palliation was achieved with a median life span at least equal to that observed with conventional therapies; (3) almost all patients had local progresion of the disease.

Based on this experience a follow-up protocol was activated in June, 1996. Under the new protocol, radiation doses to tumor and the surrounding target volume (2 cm envelop around the tumor) have been increased. To date, January 14, 1997, thirteen patients have been treated, however, it is too early to determine whether or not the increased radiation doses will result in increased life span. This study remains in progress.

Acknowledgments

This research was supported in part by the U.S. Department of Energy under Contract DE-AC02-76CH00016.

REFERENCES

1. D.N. Slatkin, A history of boron neutron capture therapy of brain tumors: postulation of a brain radiation dose tolerance limit, *Brain*. 114:1609 (1991).
2. R.G. Fairchild, Development and dosimetry of an "epithermal" neutron beam for possible use in neutron capture therapy. I. "Epithermal" neutron beam development, *Phys. Med. Biol.* 10:491 (1965).
3. H.B. Liu, R.M. Brugger, and D.C. Rorer, Enhancement of the epithermal neutron beam at the Brookhaven Medical Research Reactor, in: *Advances in Neutron Capture Therapy*, A.H. Soloway, et al., eds., Plenum Press, N.Y. (1993).
4. R.F. Barth, A.H. Soloway, and R.M. Brugger, Boron neutron capture therapy of brain tumors; past history, current status, and future potential, *Clin. Sci. Rev.* 14:534 (1996).
5. H.R. Snyder, A.J. Reedy, and W.J. Lennary, Synthesis of aromatic boronic acids. Oldehydo boronic acids and a boronic acid analog of tyrosine, *J. Am. Chem. Soc.* 80:835 (1958).
6. Y. Mishima, C. Honda, M. Ichihashi, H. Obara, J. Hiratsuka, H. Fukuda, H. Karashima, T. Kobayashi, K. Kanda, and K. Yoshino, Treatment of malignant melanoma by single thermal neutron capture therapy with melanoma-seeking ^{10}B compound, *Lancet* 12:388-389 (1989).
7. J.A. Coderre, J. Glass, R.G. Fairchild, P.L. Micca, I. Fand, and D.D. Joel, Selective delivery of boron by the melanin precursor analog *p*-boronophenylalanine to tumors other than melanoma, *Cancer Res.* 50:138 (1990).
8. J.A. Coderre, D.D. Joel, P.L. Micca, M.M. Nawrocky, and D.N. Slatkin, Control of intracerebral gliosarcomas in rats by boron neutron capture therapy with *p*-boronophenylalanine, *Radiat. Res.* 129:290 (1992).
9. J.A. Coderre, T.M. Button, P.L. Micca, C.D. Fisher, M.M. Nawrocky, and H.B. Liu, Neutron capture therapy of the 9L rat gliosarcoma using the *p*-boronophenylalanine-fructose complex, *Int. J. Rad. Oncol. Biol. Phys.* 30:643 (1994).
10. K. Yoshino, A. Suzuki, Y. Mori, H. Kanihana, C. Honda, Y. Mishima, T. Kobayashi, and K. Kanda, Improvement of solubility of *p*-boronophenylalanine by complex formation with monosaccharides, *Strahlentheropie* 165:127 (1989).
11. T.R. Lahann, C. Sills, G. Hematillake, T. Dymock, and G. Daniell, Cardiovascular toxicities associated with intravenous administration of *p*-boronophenylalanine formulations, in: *Advances in Neutron Capture Therapy*, A. Soloway, R. Barth, D. Carpenter eds, Plenum Press, New York (1993).
12. J.A. Coderre, P. Rubin, A. Freedman, J. Hansen, T. Wooding, D. Joel, and D. Gash, Selective ablation of rat brain tumors by boron neutron-capture therapy, *Int. J. Radiat. Oncol. Biol. Phys.* 28:1067 (1994).
13. J.A. Coderre, M. Makar, P.L. Micca, M.M. Nawrocky, H.B. Liu, D.D. Joel, D.N. Slatkin, and H.I. Amols, Derivations of relative biological effectiveness for the high-LET radiations produced during boron neutron capture irradiations of the 9L rat gliosarcoma *in vitro* and *in vivo*, *Int. J. Radiat. Oncol. Biol. Phys.* 27:1121 (1993).
14. R.Bergland, E. Elowitz, J. Coderre, D. Joel, and M. Chadha, A phase 1 trial of intravenous boronophenylalanine-fructose complex in patients with glioblastoma multiforme, in: *Proceeding of the Sixth International Symposium on Neutron Capture Therapy* pp. 739-745, New York, Plenum Press, 1996.
15. D.D. Joel, M. Chadha, A.D. Chanana, J.A. Coderre, E.H. Elowitz, J-O. Gebbers, H.B. Liu, P.L. Micca, M.M. Nawrocky, and D.N. Slatkin, Uptake of BPA into glioblastoma multiforme correlates with tumor cellularity, in: *Seventh International Symposium on Neutron Capture Therapy*, Elsevier Science (in press).
16. A.D. Chanana, J.A. Coderre, D.D. Joel, and D.N. Slatkin, Protocols for BNCT of glioblastoma multiforme at Brookhaven: Practical considerations, in: *Proceedings of the Seventh International Symposium on Neutron Capture Therapy*, Elsevier Science (in press).

A NEW APPROACH TO CONTRABAND DETECTION IN LUGGAGE AND CARGO [1]

Joseph J. Sredniawski

Northrop Grumman Corporation
Advanced Technology and Development Center, 1111 Stewart Avenue,
Bethpage, NY 11714-3582

ABSTRACT

A Contraband Detection System (CDS) Proof-of-Principle (POP) is nearing completion at Northrop Grumman's Advanced Technology and Development Center. It employs gamma resonance absorption to detect nitrogen or chlorine in explosives and certain forms of illegal drugs. Using tomography, 3-D images of the total density and selected element density are generated. These characteristics together may be utilized with confidence to determine if contraband is present in baggage or cargo. The CDS employs a high current (10 mA) DC electrostatic accelerator that provides a beam of protons at either 1.75 or 1.89 MeV. The beam impinges upon a target coated with ^{13}C or ^{34}S. The resultant resonant gamma rays are preferentially absorbed by ^{14}N or ^{35}Cl. Our program calls for testing of the POP CDS by late 1996. This paper presents the overall design of the CDS POP and some applications along with estimated performance.

INTRODUCTION

The potential of GRA for detection of explosives has been cited in previous work (1), (2). We have further developed the use of GRA in a system that offers the potential for detecting certain illegal drugs. The CDS POP mission is to examine the usefulness of the GRA technique with regard to explosives and drug detection, and to provide a data base to assess the practicality for use with man-portable luggage. The CDS approach may also be effective for the interrogation of medium sized cargo containers like the LD-3. Development of a high-current electrostatic accelerator for the CDS uses state-of-the-art technology that will be beneficial to other applications like radiography or medical therapies.

TECHNICAL APPROACH

Two primary characteristics that we employ to identify explosives or illegal drugs among common materials are total density, and selected element density (Fig. 1). Due to the overlap in total density of many materials, X-rays alone are not sufficient to separate contraband from common materials or materials deliberately used for concealment. Similarly if nitrogen density is used alone, false alarms may be caused by common materials containing nitrogen. The combination of nitrogen density imaging with total density imaging through the use of tomography, significantly improves the detectability and false alarm rate of GRA based CDS.

---

[1] This work was conducted under USAF Contract #F0865094C0097 funded by the Advanced Projects Research Agency.

Similarly chlorine density imaging can provide a means of detecting non-nitrogen based explosives.

Resonant gammas come off the proton target in a conical fan at 80.7 degrees for nitrogen and 82 degrees for chlorine (Fig. 2). Since these angles are so close, a common array of segmented BGO detectors is utilized over an arc of 53 degrees to provide input to an imaging subsystem. Discrimination between resonant and non-resonant gamma rays to acquire dual density images is achieved by exploiting the fact that the resonant gammas are emitted at a specific angle. Non-resonant gammas are collected off the resonant angle. The detection system is therefore required to be gamma position sensitive. A detector development program is in place and image resolution better than 5 mm has been demonstrated.

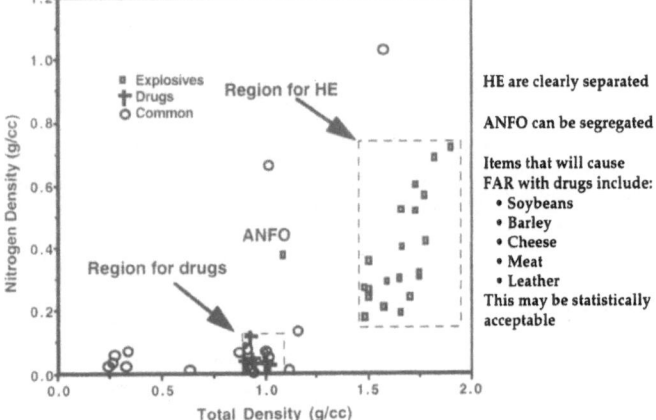

FIGURE 1. The use of partial density (nitrogen shown) together with total density can be used to discriminate threat object from common objects.

The tomographic imaging approach utilizes rotation and vertical translation of a baggage carousel while batch processing of the contents provides high throughput. A detection station consists of a proton beam target, radiation shielding, baggage handling, the detectors and data acquisition subsystem. Numerous detection stations can be employed to operate independently from each other in time share mode from a single accelerator. A single proton accelerator using a special target can simultaneously supply multiple detector arrays with the appropriate gammas to use most of the 360° gamma cone, which leads to high throughput approaching 2000 bags/hr.

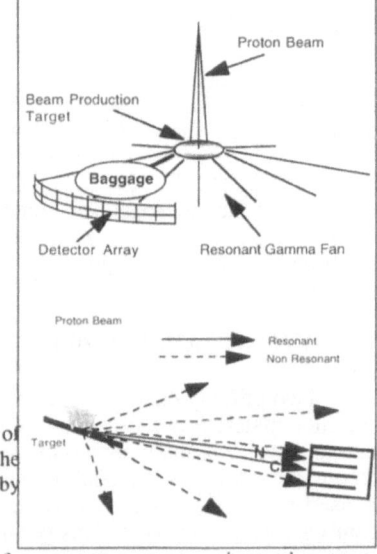

FIGURE 2. Position sensitive detection is obtained by means of segmented BGO detectors derived from PET technology. The resonant angle for chlorine is very close to that of nitrogen thereby permitting common use of the same detector array.

The narrow nuclear cross section for generation of resonant gammas gives rise to a proton beam requirement on target having small energy spread. This in turn leads to the selection of an electrostatic accelerator. For the CDS POP a tandem configuration is chosen which uses an electron stripper at the high voltage terminal and a compact high current-high voltage power supply. At the beam current densities of interest, a conventional foil stripper would be destroyed. Beam dynamics simulations for a gas stripper channel indicate that the required performance should be achievable. The issue of proper gas confinement in the stripper region to mitigate HV accelerator breakdowns has been addressed with a successful off-line test program at TRIUMF.

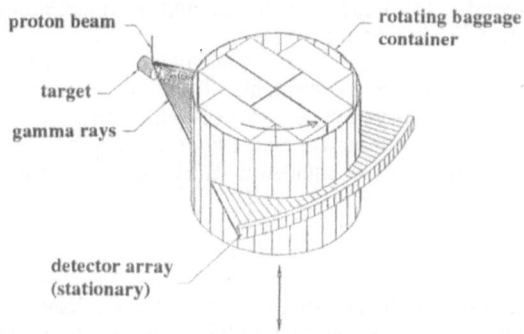

FIGURE 3. The CDS uses a tomographic inspection technique.

The need for short inspection time drives the accelerator current upward toward the survival limits of the gamma generating target. Proton currents of 10 mA or more give reasonable inspection times. The proton beam target for the GRA technique is identified as the highest risk in the system. There is limited experience about the lifetime of the target coating from constant bombardment by energetic protons; however, a parallel test program has been underway for some time at Northrop Grumman which addresses this issue. A rotating target design is employed to spread the effects of beam sputtering and heating. Care must be taken to select the correct material sub-layers so a clean gamma signal is achieved along with favorable target lifetime.

DESCRIPTION OF THE POP DEVICE

The machine is designed for maximum flexibility and ease of access for quick changes and/or modification as anticipated in a development tool rather than a machine for field use (Fig. 4). The centerline of the tandem accelerator is 108 inches above the floor.

This configuration provides adequate length in the high energy beam transport section which must be bent at an angle of 80.7 degrees for proper position of the proton beam with respect to the target surface to result in a horizontal gamma ray fan at the area of container inspection. Another choice to have the accelerator at a lower level would require that the target, the detector and container handling equipment be at elevation. We selected the former based upon the anticipation that most of the hands-on time with the POP device will be with the target, detectors and container handling. The POP shown in the figure employs a double-decked array of 88 BGO segmented detectors spanning a field of view of 53°.

FIGURE 4. Isometric simulation of CDS POP at Northrop Grumman.

Due to funding constraints on the present program, the initial POP demonstration scheduled to take place in December 1996 will use a scaled down set of detectors (single layer of 7 BGO detectors) and a smaller baggage handler than shown in the figure. Resonant imaging will also be limited to nitrogen. The proton accelerator will demonstrate full DC output, but the gamma production target will be a low-cost/low-duty factor design. The POP will demonstrate key principles that are scaleable to a fieldable CDS including; (a) high current DC tandem accelerator operation and long term stability at the required beam conditions for resonant gamma production, (b) image resolution and resonant/non-resonant gamma ray separation, and (c) basic 3-D tomographic imaging.

Fig. 5 shows the tandem accelerator arrangement with two accelerating columns extending from the high voltage center terminal. The center terminal contains two sets of triplet magnets, a gas stripper subsystem, a series of collimators and associated diagnostics. The whole assembly fits within a corona cage and sits on top of a 1 MV power supply. The external containment vessel provides an enclosure for SF_6 (dielectric) at 60 psi.

PERFORMANCE

The counter-drug mission differs significantly from explosives detection in that the volumes of contraband involved are usually larger when compared to high explosives, relaxing the need for high resolution. The larger volume of contraband associated with drug trafficking (typically ≥ 1 kilo) permits integration of the 3-D scan data over larger slices, thereby enhancing the signal to noise and permitting image segmentation to improve the statistical probability of detection. On the other hand the larger amount of nitrogen in high explosives makes detection less difficult.

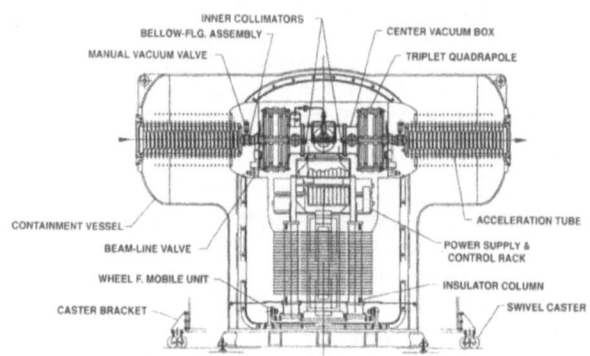

FIGURE 5. Side view of the tandem accelerator arrangement.

Simulations of expected performance have been made with a fairly sophisticated model that includes background clutter material as well as a mix of potential threats. To be realistic some of the non threat background materials include nitrogen. For the model, a mix of 70% clothing, 25% dense objects and 5% voids was assumed. Predicted performance is shown in Fig's. 6 and 7 for explosives and drugs respectively. All data is based upon a 5 mm scan slice thickness.

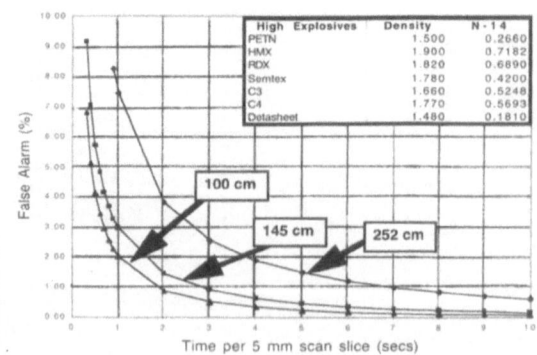

High Explosives	Density	N-14
PETN	1.500	0.2660
HMX	1.900	0.7182
RDX	1.820	0.6890
Semtex	1.780	0.4200
C3	1.660	0.5248
C4	1.770	0.5693
Detasheet	1.480	0.1810

FIGURE 6. Using nitrogen imaging the false alarm probability as a function of scan rate for a sensitivity of 450g of thin sheet high explosive in various sized containers is shown. The probability of detection is fixed at 90%.

The CDS performs most efficiently when the penetrating power of the 9 MeV gammas is employed to perform batch processing of baggage or inspection of cargo containers. For small or loose baggage, a carrousel can be sized for an optimum volume to maximize throughput and still maintain sufficient transmission of gamma rays for high probability of detection. For packing density of 85% the CDS could process up to 430 bags/hr as shown in Table 1 based upon the statistically average bag of 2.27 ft^3. Chlorine based imaging is about a factor of 20 slower.

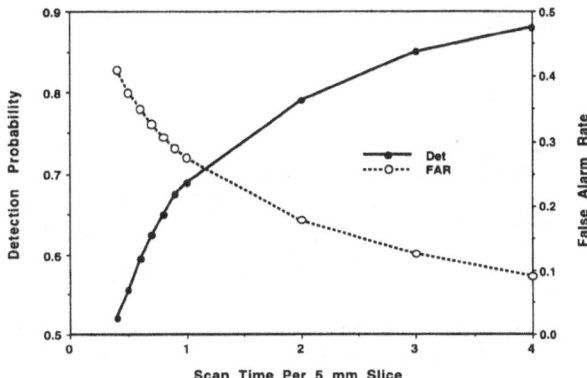

FIGURE 7. Estimated detection performance for a 1 kilo quantity of cocaine concealed in a loaded LD-3 aircraft container. Performance is based upon nitrogen imaging.

TABLE 1. Characteristics of a baggage inspection type CDS

CHARACTERISTIC	VALUE	
	Nitrogen	Chlorine
Element Detected		
Nuclear Reaction	$^{13}C(p,\gamma)^{14}N$	$^{34}S(p,\gamma)^{35}Cl$
Target Type	segmented surface	
Beam Current (mA)	≥ 10	10/20
Beam Energy (MeV)	1.75	1.89
Energy Spread (keV)	25	12
Detector FOV (degrees)	53	53
Container Dia. (cm)	145	145
Container RPM	60	6
Detector Height (cm)	10	10
Detector Radius (m)	2.14	2.14
Resolution (mm)	5	50
Est. Q (bags/hr)	430	25/50
Sensitivity (kg)	0.5	4.0
Detection Probability	0.9	0.7

Conceptual designs of deployed CDS machines are included in Figures 8 and 9.

CONCLUSIONS

In contrast to what has been previously published about the viability of GRA (3), we have developed a suitable approach that promises to provide excellent performance in the detection of explosives and drugs. Although there are some areas of the CDS POP that may not be completed to the full potential due to limited funding at this time, the planned demonstration will suffice to show whether the GRA approach is practical for either or both drug and explosives detection. The POP device will benchmark our models and facilitate prediction of fieldable CDS performance. Development of the high current tandem accelerator will have applications other than CDS.

ACKNOWLEDGMENTS

The work conducted on this project is the result of a team consisting of Northrop Grumman, TRIUMF and Scientific Innovations, Inc. I sincerely thank all of our team members for their respective contributions.

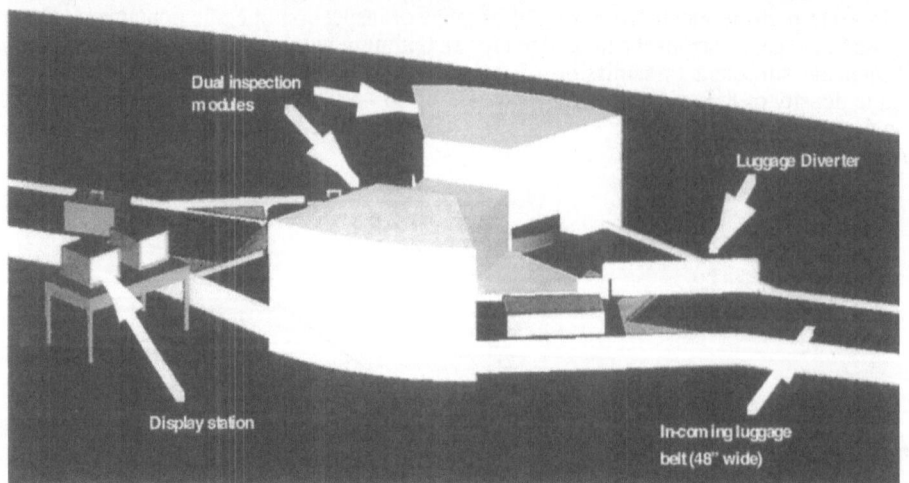

FIGURE 8. Concept for a dual inspection station installation at an airport facility with individual baggage transport. The accelerator would be located remotely, possibly within the basement of the building.

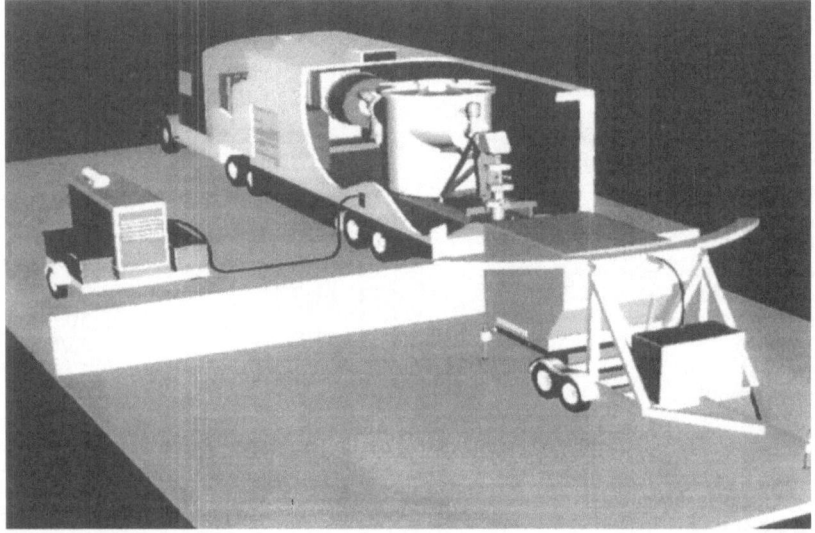

FIGURE 9. Concept for a transportable CDS to inspect LD-3 containers. An advanced accelerator design is employed for compactness.

REFERENCES

1. Gozani, T., *"Nuclear Based Techniques for Cargo Inspection* - A Review", Proceedings of Contraband and Cargo Inspection Technology International Symposium, pp. 9-19, October, 1992.
2. Vartsky, D., et al., *"A Method for Detection of Explosives Based on Nuclear Resonance Absorption of Gamma Rays in 14N"*, Nuclear Instruments & Methods in Physics Research, Section A 348, pp. 688-691, 1994,.
3. Loveman, R., *"Nuclear Resonance Techniques for Drug and Explosive Inspection in Luggage"*, Proceedings Counterdrug Law Enforcement: Applied Technology for Improved Operational Effectiveness International Technology Symposium, Part 1, October 24-27, 1995.

THE FUTURE OF NUCLEAR POWER

Juan Eibenschutz

Subdirector de Distribución y Comercialización
Luz y Fuerza del Centro
México, D.F.

Nuclear power became a major player in the energy balance of the world thanks to the research and development programs of several countries. Furthermore, the nuclear power industry flourished with subsidies from governments, in the days when energy was considered a strategic priority by most countries, because it offered a valid alternative to the existing energy sources, and also since some countries were interested in its weapons related capabilities.

Several countries constructed nuclear power plants with the two objectives of producing weapon materials and energy. Others selected designs that were highly impractical for the production of weapon materials and were only efficient as power plants. Some of the power reactors, even if constructed for war purposes, (like submarines) were not useful to produce weapons.

During the fifties nuclear power became economically competitive, even if some opponents argued that there was not true competition, since the original government subsidies that financed the development of related technologies were really the reason for achieving competitiveness. But government subsidies are the main support for most of the technological developments that have taken place in the world, not only in the energy field, but in practically every one of the activities that have made progress possible.

Had it not been for its economic merits, nuclear power would have never achieved its present level of penetration; those merits are still there, however, several things have changed over the last decades.

First of all, nuclear power was born under the diabolic spell of the atomic bomb, which laid dormant in the minds of the general public, until accidents, that were heralded as impossible by the nuclear industry, did happen, awakening the fears and providing fresh arguments to opponents. So much so that the nuclear issue became very important in political debates, and as in many other cases, difficult to defend, since politicians find it almost impossible to adopt a pro-nuclear platform when there is a capital sin against it in the minds of voters.

As a second point, related to the first one, regulatory bodies overreacted to the accidents and created situations of uncertainty, that made it very expensive and risky to

Technology for Global Economic and Environmental Survival and Prosperity
Edited by Kursunoglu *et al.*, Plenum Press, New York, 1997

263

invest in nuclear power. On the one hand, some of the regulatory changes required costly retrofits, on the other, the concept of inherent safety was proposed by members of the nuclear community.

Inherent safety was immediately adopted as specific to nuclear power, generating still more concerns, since the present reactors were automatically labeled as inherently unsafe. The question is: is there any man made system that can be considered absolutely safe?

Obviously no. Every man made machine, process, facility, device, etc. is subject to failure; even naturally occurring systems, like the earth itself, are subject to failure. What counts, and what society does, is to design for the lowest probability of failure. And in the case of natural structures, to assess probabilities of failure and either get away from them or design to cope with failures.

Two stories illustrate the situation. The accident rate of the nuclear industry has been extremely low, therefore, society is so unaccustomed to nuclear accidents that Three Mile Island practically stopped further development of nuclear power in many countries; the DC-10 Chicago accident stopped flights of that type of aircraft only for a few weeks. No one was physically harmed by Three Mile Island, more than two hundred people died in the aircraft accident !

Air conditioning (HVAC) in nuclear power plants is required to keep components at specified operating ambient temperature and humidity. The HVAC systems employed in nuclear power plants are conventional, but require large volumes of space, since they operate practically at atmospheric conditions. Ships, aircraft and several other installations employ high velocity HVAC schemes, that require considerably less space; when questioned about this costly inconsistency, architect engineering firms, and nuclear power systems designers, systematically answer: conventional HVAC is already licensed !

There are many more stories about the discrepancies between dealing with nuclear power and with other energy forms or other types of industry. Nevertheless, the fact is that the differences between the perception of risk and its quantitative assessment, although present in many fields, are specially important in the case of nuclear related issues.

Inspite of these phenomena of "unfairness" nuclear power is one of the best energy sources, particularly from the environmental perspective. Nuclear power operates with closed-loop processes. All the fuel cycle steps take place in enclosed systems, under suitable containment conditions, that prevent potentially harmful substances from accessing into the environment. Even under accident conditions the potential damages of nuclear power systems are much less than those of other industries; the basic difference lies in the concept of "defense in depth," that applies to design, construction and operation of nuclear facilities, including postulating accident scenarios to demonstrate that even during accident situations, damage is mitigated and kept to minimal levels.

This paper however, does not pretend to add much to the considerable material that has been published on the subject. The intention is rather to point out that the present trends in the power industry are not favoring nuclear power and may further endanger the future of overall energy supply.

As pointed out earlier, government subsidies were fundamental in the global deployment of nuclear power. A fact not unique to nuclear power, but in this case clearly attributable to the strategic value that most countries assigned to energy.

With deregulation, globalization, privatization and the creation of conditions of "forced competition" in the power industry, the long term issues and the strategic nature of energy are in the process of abandonment in most countries, and in some, energy supply has already been left solely to market forces.

Nuclear power is particularly ill suited to compete under traditional market conditions. A clear example is the issue of stranded investments in nuclear power plants, presently very much "en vogue" in the U.S.A.

Traditional market conditions apply mostly to the short term. If for example, the prices of fuels like coal, gas, or oil change, the relative merits of power plants fueled with these fuels change, and some power plants become stranded investments, or even go bankrupt.

Capacity additions in energy systems, and specially in power systems, are made considering the overall system as well as its growth forecasts. Rate payers are charged for a security of supply, and regulators were responsible to assure that conditions for fair play were warranted. Decisions for capacity additions, under these rules, require power system modeling, and forecasting of the future behavior of the main parameters.

Since electricity cannot be stored, errors leading to future capacity deficits can be orders of magnitude more expensive than the economic effects of excess capacities. This is why assessing the individual economic performance of investments in power plants belonging to interconnected power systems makes little sense.

Energy supply is clearly not a short term business. Market situations may promote wonderful deals for some and dramatic losses for others, but the market economy cannot possibly assure that there will be no energy shortages. As a matter of fact, energy shortages as well as power shortages, may make lots of money for the operators.

Unless there is some external action, one of the best business mechanisms to raise power prices, and therefore make more money without further investments, is to let demand outgrow supply.

Letting the market forces act in the power supply business, as the sole mechanism, may lead to future insecurity of supply. Undoubtedly, competition is one of the best price regulation mechanisms, but in energy supply, prices are but one of the factors requiring regulation.

It is obvious that if the present policies of competition in power supply fail to provide for the required generating capacity, governments will step in and take appropriate actions to provide the required capacities, either by directly installing additional power plant, or by applying suitable regulatory reforms.

The point is that under the present market policies, additional barriers are being created against nuclear power, the following discussion attempts to show why.

Investors in power plant seek to lower their exposures by minimizing investments, since the market system allows to charge for changes in fuel prices. Nuclear power plants have high investment costs and low fuel costs, therefore, the market policies will limit their use, even if they compete on economic grounds.

Energy security becomes irrelevant under market conditions. Most of the energy resources being exploited are finite, even if the experience during the last few decades seems to indicate otherwise. In this case there is the emergency fuels sharing mechanism for the member countries of the International Energy Agency, which could mitigate the effects of shortages. Even if it would make more sense to promote a more balanced energy supply, because of the time periods required for the deployment of nuclear power, fuel shortages could not be compensated, practically, with nuclear power.

One of the characteristics of power systems is that they benefit from economies of scale. As a matter of fact, huge interconnected power systems supply electricity to the largest and wealthiest countries around the world, reliably and economically. Nuclear power provides roughly the same proportion of energy supplied by hydroelectric power, another source being opposed by environmentalists on the grounds of their local effects on the ecology and the climate. Hydroelectric and nuclear power are not the favorites of independent power producers because aside from the relatively high levels of financing they require, they are faced with very stringent regulatory requirements. The question is: can we afford to ignore these energy sources?

When the market economy dictates, it really does rule. Social and other considerations are pushed aside in favor of profits. The importance of reduced schedules for the execution of projects increases dramatically, and therefore, the need to comply with time consuming regulatory processes can rule out certain solutions, even if they dominate on economic and other grounds.

The present situation of globalization and open markets, together with the extremely effective mass communication systems, are promoting not only the market economy as the global solution, but also the globalization of concerns, regardless of whether they really apply to the conditions of every country.

The following discussion attempts to illustrate this.

Climate change has become a major issue in energy policy making, as part of a broader energy-environment culture. Paradoxically, nuclear power, the environmentally more benign energy source, is considered environmentally undesirable. For example, radioactive waste storage is opposed on subjective grounds, since technically it has been demonstrated that sound solutions are economically feasible. As another example, decommissioning costs for nuclear power plants are being made part of operating expenses. Who talks about decommissioning of hydro power facilities, and restoration of dam and reservoir sites to their original status. Or, has anyone worried about restoring coal mines?

Then there is natural gas. Developments in materials technologies, together with better designs, have lead to a new generation of gas turbines, that allow for enormous increments in power plant efficiencies in combined cycle plants. Furthermore, gas firing produces much less CO_2 than coal per unit of electrical energy, thereby giving gas fired power plants environmental advantages that make them still more attractive.

Nevertheless, gas firing is not without environmental effects. Aside from the combustion gases themselves, methane, much more effective as a greenhouse gas, will be released to the atmosphere in increased amounts, as a result of much higher consumption levels.

But the main cause of concern should be economical. With present prices, combined cycle gas fired power plants are the cheapest source of electricity. However, as demand for gas grows, transportation from remote sources will be required, and this will make nuclear power competitive, with an ample margin. The problem is that decisions are not taking into consideration neither price increases, nor environmental costs. By the way, the only power source where practically all the economic externalities are internalized is nuclear power.

Although renewable energy sources have been used since the appearance of man, there seems to be presently a sort of crusade in favor of renewable and "soft" energies. This in itself is not necessarily bad, since diversification of energy resources is important for many reasons. The problem is that too many people believe that solar, geothermal, biomass, wind, OTEC, etc. can provide the energy required by society, in enough quantities to substitute, not only for nuclear energy, but even for all types of fossil fuels.

Since ultimately policies, and particularly energy policies, have to conform to the will of the people, the risk of converting unfeasible wishful thinking into policies becomes a distinct possibility. Regardless of the fact that in the foreseeable future it is not possible to do without conventional energy sources (including nuclear), some countries have already enforced policies to ban all energy forms that are not solar or renewable, for their future energy supplies.

As an example it may be relevant to consider aeolic energy (wind power). Technology for this type of power source has achieved economic competitiveness, and many believe that the reason why this clean, environmentally benign, and cheap power source is not fully deployed, to substitute for coal, oil and gas fired power plants, and of course for nuclear power plants, is that there are too many vested interests against wind power. Apparently very few people are aware of the fact that even in sites where wind velocities are high and practically constant over time, wind energy does not provide firm power, and therefore this source can only be used when guaranteed availability is not a requirement. This means that normal power systems can only use aeolic energy if production costs are lower than the fuel costs of the marginal plants in the system.

Joint implementation, a mechanism devised by industrialized countries to comply with environmental targets for greenhouse gas emissions, now under test by several countries, promotes energy efficiency and also the deployment of renewable energy sources. It is interesting to note that even if the credits are not yet being traded and that nuclear power would be the most effective greenhouse gas source of substitution, promoters of joint implementation freeze when this fact is put before them.

Climate change is receiving great attention in many quarters around the world. Many politicians argue that the effects of man induced climate change may be of such magnitude that it is irresponsible not to do everything possible to curtail emissions of greenhouse gases. Nevertheless, this same politicians are not even willing to discuss the merits of nuclear power.

The arguments in favor of nuclear power are considered overwhelming. . . . . unfortunately, mainly by the nuclear community itself. The preaching to the believers, a phenomenon not unique to nuclear, reflects the bitterness of those who think that their truth is being treated unfairly.

Galileo was right, but it took society a long time to recognize it. Unless there are drastic changes in the world, either societal, or scientific, the future seems very, very difficult without nuclear power.

What may be called a positive scenario: the scientific breakthrough of direct conversion from the energy contained in nuclei into useful energy, without having to go through thermal energy, would assure the future supplies practically forever.

But a horror scenario: the drastic reduction of world population as a result of climate change, could also make for a sustainable future.

Let us go for the first scenario.

CHAPTER VI

KEY ISSUES IN NUCLEAR POWER: A WORLD COMMISSION ON NUCLEAR ENERGY

A WORLD COMMISSION ON NUCLEAR ENERGY: REASONS & PURPOSE

Peter Beck

Royal Institute of International Affairs
10, St. James's Square
London SW1Y 4LE

INTRODUCTION

Beliefs about the future of nuclear energy have changed drastically over the last 20 years. It was seen as the main hope of mankind for cheap and abundant energy during the mid-1970s, but today there are doubts in many quarters whether this energy form can survive in the light of its unresolved problems, unfavourable economics and unpopularity. Although the nuclear industry strongly challenges the basis of this change in perception, there is little doubt that public fear of nuclear power is high and that proposals for new nuclear facilities have run into political barriers in most OECD countries. A book, published by the Royal Institute of International Affairs[1] in London, attempted to analyse the reasons why these changes came about and how these could affect the future development of this energy form. The concept for a World Commission on the future of nuclear energy arose from the conclusions of that book.

This paper considers these conclusions, develops the case for the type of study suggested for the Commission and finally looks at its possible structure and mode of working.

THE FUTURE OF NUCLEAR ENERGY - AN ANALYSIS

1. Judgement - the Achilles Heel of Forecasting

The future of nuclear power will depend on a vast variety of factors, from its economics relative to other energy forms, growth of energy demand and how such demand could be met, to environmental issues, technological developments, safety, security, connection with weapons proliferation and the political fallout from the public fear of radioactivity. Because of the long term nature of nuclear investment, it is necessary to look 30 to 50 years ahead,

Technology for Global Economic and Environmental Survival and Prosperity
Edited by Kursunoglu *et al.*, Plenum Press, New York, 1997

271

even if the purpose of the analysis is to assist decisions which have to be taken within the next few years.

Unfortunately, there are no 'facts' about the future. To assess all the factors mentioned, is a matter of judgement of what may reasonably happen between now and, say, 2050. From experience of the past, the range of possibilities is so wide, that virtually any conclusion becomes possible. And that, indeed, is what is happening.

The nuclear industry and its followers argue plausibly that the energy form can make use of an energy resource base which (with use of Pu recycle and FBRs) is perhaps twice the size of all fossil fuel resources; that it can assist the battle against global warming; that when judged on a 'level playing field' its economics is competitive; and that its safety record is no worse than those of other energy forms. Antagonists argue equally plausibly that nuclear energy is too complex, too expensive, too dangerous and too proliferation prone, especially if Pu recycle on a large scale were to be considered. Furthermore, they believe that the world can get by without nuclear power, even if use of fossil fuel were to be curtailed.

Both parties are convinced that their view of the future is correct and find it difficult to understand how the opposition can truly believe what it is saying; there is thus mistrust between the parties with accusations of ulterior motives being traded. As a result, there is little dialogue between the differing viewpoints and the factions talk past each other. With present-day mistrust of governments and large organisations, the public has tended to side with the environmental lobby, which in many countries is the backbone of the resistance against nuclear power. Because of that, expansion of nuclear power has come to a virtual halt in most democratic countries with only France and, at least until recently, Japan the exception.

It has to be noted that in addition to the pro- and anti-nuclear factions, there is a third party. It believes that nuclear power should have a major future, but has doubts whether, in a nuclear industry perhaps five to ten times larger than now, today's technology is safe enough and the fuel cycle with Pu recycle and FBRs sufficiently proliferation proof to make such expansion acceptable to the public.[ii] It also believes that technological advances have shown that better alternatives exist, but considerable effort and funds would be needed to prove them commercially[iii]. The judgement of the industry strongly disagrees with such views, believing that continuing development of existing processes provides adequate safety and security. The anti-nuclear lobbies consider different processes and fuel cycles to make little difference because to many it is a matter of conviction that nuclear energy is fundamentally flawed.

In summary, one of the disturbing conclusions of the book is that prediction of the future of nuclear energy is bedevilled by a vast range of judgmental differences between various lobbies. Some of these differences may be due to bias and prejudice, but others are due to the wide range of uncertainties about the future which no amount of additional investigation can narrow. With little dialogue between the parties and most studies and conferences aimed at convincing the convinced, there is little chance of achieving agreement or even understanding of each others' point of view between the various parties. Yet, decisions have to be taken, but with strong pressure groups providing opposing advice, many countries settle for unrealistic decisions or a lack of actions which, if continued, will lead to their slow abandonment of nuclear power, while leaving major safety and security problems to future generations. The proposed Commission can be a means of breaking the present stale-mate by taking evidence from all sides, catalysing debates between the various lobbies and so provide better understanding of what is or is not achievable.

2. Will there be a Need for Nuclear Power?

The answer to this question largely depends on a number of uncertainties which include the future world electricity demand, availability of alternative fuel resources and the issue of global warming. There are very divergent views about such subjects. During the early 1970s high estimates of future energy demand and indications that fossil fuel resources might not be adequate to meet them, led to the view that rapid expansion of nuclear energy was essential to avoid future energy crises. Far lower growth of world energy demand during the 1980s, more optimistic assessments of fossil fuel resources and greater interest in renewable energy sources now indicate that, at least for many decades to come, there could be adequate energy resources even without nuclear playing a major role.

That view, however, did not take the problem of global warming into account, which, if confirmed as a major threat, could lead to the need to constrain CO_2 emission and therefore limit the use of fossil fuel. At the time, 1992, of the Rio Earth Summit conference, the effect of global warming had not been scientifically fully confirmed and the conference agreed to take only such precautionary measures as are 'cost-effective' and 'comprehensive'. If nuclear energy were to be seen as economic and, politically acceptable, there can be little doubt that continued investment in nuclear power would fit this 'precautionary principle'.

These changes in perception over the last two decades about the long term world energy supply/demand picture indicates the great difficulty of making such forward projections. Looking now 30-40 years ahead opens up even greater uncertainty. On the demand side, world population is to increase substantially, with nearly all increase to be in the developing world. Furthermore, the proportion of the LDCs' economic activity in the world may well increase from well below 50% today to 70% or even more. This implies that the developing countries will need more energy, but how much more is a matter of much debate and no conclusions. Environmental lobbies believe that it should be possible for the developing world to become far richer without following the west in using ever-larger amounts of energy, while others have grave doubts whether there is either the means or the will to achieve this. On the supply side, there is the question of global warming and its effect on use of fossil fuel and very divergent opinions about the potential of renewable energy.

The rational response to uncertainty is flexibility. Nuclear energy is one of only two means of electric power production which do not produce CO_2 (the other being some forms of renewable energy). The choice whether large scale use of one, both or neither will be needed to balance the world's future energy system cannot be made at the present time, but to maintain flexibility would appear to imply that both continue to be options for strong world-wide expansion, should the need arise.

If the problem of global warming is confirmed, relegation of nuclear energy to a minor role is something the world may live to regret. But what does 'keeping options open' mean and can it realistically be done? That is a question for the Commission.

3. Safety, Radiation and Public Attitudes

The possibility of releasing radioactivity into the environment and fears about catastrophic release, say by a reactor melt-down, became an early public concern. The original civil programme in the USA went to great lengths to reassure the public about the safety of its designs, but the industry's assurances were rather dented by incidents in the military programmes of the nuclear weapons states and by events, such as Three Mile Island during the 1970s and finally by Chernobyl. Whether justified or not, these incidents, plus a general fear of radiation and its effects, caused public opinion to become risk-averse about nuclear power. This increased local resistance to siting nuclear facilities or dumps and caused severe delays in obtaining planning permission for new plants, at least in democratic countries. The industry and governments attempted to show that this mistrust was misplaced and they acted by increasing safety provisions and by attempting to define safe limits of radiation doses; but to little avail.

Research[iv] has shown that individual attitudes, which are the basis of public attitudes, appear to be formed by aggregating, perhaps subconsciously, a number of beliefs about risks and benefits; complex statistical arguments and comparisons employed by the industry play little part in such an analysis which tended to find nuclear energy too risky for the perceived benefits. When it came to believing governments and the industry or the environmental lobbies, the public in many countries chose the latter. Led by the environmental pressure groups, public opinion was especially concerned about two fields.

Reactor Safety: Although the industry seems justified to argue that, excluding Chernobyl, the injuries record of reactors is good, there are strong views, some from inside the industry, arguing that present designs are too dependent on engineered safety and on high operating and maintenance standards[v]. As long as plants are under the eye of a effective regulatory regime, such designs might indeed be safe enough., However, if many more reactors come into operation, some in countries with no such regime, possibly under pressure to save on capital cost and not used to admitting mistakes, what then?

Back-End of the Fuel Cycle: That, today, is still an unresolved issue. The argument whether to reprocess or not to reprocess spent reactor fuel is splitting the industry and governments. That does not assist public confidence; nor does the fact that permanent disposal facilities for HLW, which in some countries were originally to be ready in the 1980s may not be able to receive material until well into the next century, while other countries have not even commenced planning what to do with such materials. Indeed, there are voices[vi] who wonder whether the present plans for deep underground disposal are not flawed. Finally, there are worries about the safety of reprocessing plants, especially if the present designs continue to be utilised in many countries as part of a major expansion of nuclear energy.

One factor feeding public mistrust is the fact that there are at present no global and mandatory monitoring systems in place to ensure that new designs of nuclear plants, storage facilities and means of transportation are safe and that operating and maintenance standards of all such facilities are of the required high level at all times., A danger is therefore seen, that short cuts to reduce cost could be taken, or that discipline becomes lax which might eventually lead to another Chernobyl. Accidents in other industries, which pride themselves in having very high standards, show such fears to be well justified; Piper Alpha in oil, Bhopal and Sevesso in chemicals spring to mind.

The fact that radioactive clouds do not respect national frontiers has further increased public fears because feeling safe has become a matter of not just trusting one's own government and industry, but also that of one's neighbours. If one country believes that another has unsafe facilities, but refuses to deal with this problem, serious conflicts could well arise.

With all these problems and disagreements in mind, it is not surprising that the anti-nuclear lobbies are alive and strong in most countries where dissent is tolerated. Keeping the nuclear option could mean that public attitudes have to change. How could that be achieved? - Another question for the Commission.

4. The Weapons Connection

The issue of weapons proliferation and the spread of the peaceful use of nuclear energy revolves around the question how far involvement in the latter could assist a country or an organisation to acquire sufficient fissile materials for one or more nuclear weapons. The industry and many governments presume this risk to be low, but others, and especially the US Government, consider that nuclear power with widespread Pu recycle could cause dangers. As a result, US official policy is against reprocessing within the civil nuclear programme, whilst many other countries, especially France, Japan and Russia, take a different line. The environmentalist lobby seems to be on the side of the USA.

Notwithstanding the success of the NPT regime, worry about proliferation has increased during the last few years. Concerns are concentrated into the following areas:

I. **Doubts about the effectiveness of IAEA Surveillance**. Recent experience with Iraq and N. Korea showed that even though a country is party to the NPT and under surveillance by the IAEA, it can still go ahead with the clandestine development of nuclear weapons capability. The NPT regime can make it more difficult for a country to acquire nuclear weapons, but it cannot stop it, especially as there are always 'rogue' organisations, even within major OECD countries, which, for a requisite reward, are willing to provide illegitimate help.

II. **The Threat of Terrorism**. While previously the main proliferation threat was seen to come from states with nuclear weapons ambition, to this has now been added greater public awareness of the terrorist threat, whereby, with or without connivance by a state, nuclear material might be seized by clandestine organisations and forged into a weapon, which, however crude and unreliable, could become a potent threat in the hands of terrorists.

III. **Nuclear Disarmament**. Perversely, the dangers of proliferation have been underlined by the partial nuclear disarmament resulting from the START treaties, which should release some 200 t of Pu from military programmes. Whether used in MOX fuel, or disposed of by deep burial, or by a number of other means under study in the USA and Russia, all the alternatives will take considerable time to come to fruition and in the meanwhile such weapons grade Pu has to be seen as the ideal material for proliferators and has to be safeguarded accordingly.

All these factors mean that proliferation dangers have now gained far more political attention than some 20 years ago. This raises some difficult long term issues, e.g. whether large scale recycle of Pu involving many countries, such as envisaged in the 'plutonium economy' can ever be made sufficiently secure. That, in turn, throws doubts on the future of a fuel cycle which includes FBRs and reprocessing of spent fuel, the cycle still preferred by much of the world's nuclear industry. But are there more secure alternatives?

5. Economics and Energy Security

The relative economics of nuclear power versus other energy sources is one area where detailed calculations can be made, but the subject is also an example of the limited worth of such calculations when considering the longer term. By the judicious use of rational and logical assumptions, one can get any answer one wishes.

Under present circumstances, with ample availability of fossil fuel and relatively low prices, far higher capital cost for nuclear stations, the ill-defined destination of spent fuel and nuclear waste and a wide range of cost estimates for dismantling and decontaminating redundant plant, nuclear power seems to be at a considerable economic disadvantage. However, nuclear protagonists can argue that conditions within a decade or two may be quite different, with far higher fossil fuel prices, CO_2 penalties and, as more experience is gained, lower cost for new nuclear plant and better definition and lower costs than today's assumptions for waste and dismantling. The opposition can argue the opposite, showing that the industry had always expected lower costs, but rarely achieved them, that the cost reduction of some renewables would make these energy forms competitive with fossil fuel, which in turn could keep the level of fossil fuel prices low.

Today's disadvantage of nuclear energy is no more a pointer for the future, than the perceived advantages seen in the 1970s when oil prices were assumed to reach 100$/bbl by the end of the century and combined cycle gas technology had not yet been developed. Estimates of longer term economics are, thus, a very poor guide for longer term policy, but they are, nevertheless, commonly used. But what alternatives are there?

One school maintains that considerations of strategy within an individual country provides more robust choices. Countries, such as France and Japan, with few indigenous energy resources, took the strategic decision during the 1970s that the expansion of nuclear power would give them far greater energy security than dependence on oil and gas imports.

The policy of favouring nuclear energy for strategic reasons appears to be spreading to a number of E. Asian countries whose economies need rapid expansion of electric power. Their interest in nuclear power seems to indicate that in an uncertain world they are willing to pay a premium for reducing dependence of their economies on imported fuel, or, if they have fossil fuel resources, they believe that these are better used to bring in foreign exchange.

Of course, the opponents of the use of such strategic decision making argue that the past has shown that strategic choices can be as wrong as those based on economics, but that the former can be far more damaging. So how should such decisions be taken, especially in a field, as political and complex as nuclear energy?

SCENARIOS FOR THE FUTURE

With the above findings in mind, one can consider different future developments of nuclear power. Three distinct ones were chosen for that purpose:
1. Continuation of today's situation with little change in generating capacity.
2. A positive policy to phase out the industry altogether.
3. Major expansion.

Each case has to be examined for realism - under what conditions would the particular scenario become realistic.

1. **Continuation of today's situation.** Present nuclear capacity of some 350 GWe delivers some 17% of the world's electricity. With the economic case for nuclear power doubtful, there can be little realistic hope for major expansion of nuclear energy within the OECD. However, there are strong signs that a number of developing countries (especially China) are aiming at a rapid expansion of nuclear power and in some of these countries public opinion is unlikely to resist such expansion. In aggregate then, nuclear power may well continue around present levels, perhaps by reaching a steady state when expansion in LDCs matches closures of plants in OECD countries which reach the end of their life span.

The overall effect of such a development would be for the proportion of the world's electricity generated by nuclear plants to fall from today's figure of 17%, possibly to below 10% by 2020 and as low as 5% by 2050, with the majority of nuclear capacity slowly moving from OECD to the developing world. Under such a scenario the risks and costs of developing new technology will be seen as too great so that one cannot expect any major changes to existing techniques, whether seen as safe or not. Furthermore, without the skill-pool and financial capacity of a healthy and expanding industry, the questions of what to do with waste and the many other concerns may well be hard to resolve

As matters stand, the case for today's level of capacity to continue is quite strong, but such an industry would be slowly dying and be an unlikely springboard for major expansion.

2. Phase-out of the industry. This is the scenario devoutly hoped for by the anti-nuclear lobbies. However, to make this option realistic, a number of questions would have to be considered and high on the list would be the question of the destination, safety and security of waste, spent fuel and of decommissioning. Furthermore, under what conditions could one foresee acceptance by all relevant nations of a total phase-out throughout the World?

Although some rich countries may take such a decision, the cost of doing so by the early closure of nuclear plants is immense and unlikely to be afforded by most countries. One lesson of Chernobyl is that even another major accident may not cause the abandonment of nuclear power; when it comes to choosing between living with the possibility of a major accident or the certainty of no heating next winter, experience in the Ukraine has shown that few of the public will choose the latter.

Within the foreseeable future this alternative appears quite unrealistic, even should there be another 'Chernobyl'.

3. A major expansion. This might imply an expansion to deliver, say, 25% of the world's electricity by 2050. There would have to be 4 -5 times today's reactor capacity and it may well be that, without change in technology, Pu recycle would have to be practised by that date or soon after, with reprocessing plants and FBRs scattered throughout the world. However, if nuclear power is to be maintained as an option for expansion, it has to become politically and economically acceptable and it must be suited for the developing world, where most of the increase of demand for power will take place. The questions to be looked at for achieving this and to make rapid expansion possible should the need arise, could be:

a). Is it possible to make the design of reactors safer than today's designs, especially under conditions of mal-operation or lack of adequate maintenance.?

b) Can the economics of nuclear power be improved, especially as regards the capital cost per power produced and for smaller plants, more suited for the Developing World?

c). Is it possible to find effective solutions for the secure and acceptable disposal of spent fuel? If that includes reprocessing to extract Plutonium for re-use in fresh fuel, means would also have to be found to ensure that such activity will not increase the risk of nuclear proliferation. Ways of destroying or storing radioactive waste and Plutonium, the latter to be safe for thousands of years and acceptable to the public, would also have to be developed.

d) Means would have to be found to assure the public that internationally accepted systems for setting standards and monitoring will be in place to ensure nuclear safety and security.

Technological developments during the last decade have shown ways of possibly meeting the first three conditions, but so far only in laboratories and far more work would have to done before these could be ready for commercial application. To achieve this, would demand a considerable effort, culminating with testing on a commercial scale; that might take 20 - 30 years and cost many billions of dollars. Although such a delay may be acceptable, and might be utilised to build the international institutions needed for questions 3. and 4, there must be grave doubts whether funds for such a programme will be forthcoming. Governments are trying to leave more and more of such work to the private sector, but bearing in mind the long time scale and the great uncertainty whether or not even these new developments will be sufficient to make nuclear power acceptable, that sector is likely to find this work too risky.

It will be seen that of the three scenarios, only the third offers the possibility of an option for major expansion, but only if considerable efforts and funds are made available to make nuclear technology more acceptable. Even then, there is the risk that such work will not succeed. Examination of the other two scenarios do not show these to be easy options and there is the likelihood that both will also require substantial development and funds to ensure that the industry does not leave unsafe materials and structures to haunt future generations. Thus, all the options have their risks and difficulties; each is controversial. With so many entrenched attitudes, debate about them is at an impasse which has to be broken before there can be realistic assessments which of the many leads to follow. An examination done by a recognised neutral body, such as the proposed Commission, might be able to achieve this.

THE WORLD COMMISSION - ITS PURPOSE AND STRUCTURE

The remit for the Commission would be to provide an informed and unbiased analysis of the range of issues, and their inter-relationships, which may have a significant effect on the future of nuclear energy. The more important ones, which include some already mentioned in this paper, are:
* Safety and its various aspects.
* Security and Proliferation, especially the relationship, if any, with the spread of nuclear power.
* Essential regulatory requirements.
* Disposition of Nuclear Waste
* The environmental advantages and disadvantages of nuclear power in relation to other energy forms.
* Energy security, world-wide and of countries.
* Economics vis a vis other energy forms.
* The feasibility and economics of smaller-scale nuclear power plants and simpler fuel cycles.
* Technology, present and new developments.
* Public opinion, its formation, strength and effect on nuclear power.

These issues have to be looked at in the context of differing economic and political interests and structures of countries and regions and of inter-governmental regimes addressing global safety, security and pollution. The Commission would, from a *strictly neutral stand-point*, take evidence from all sides, consider these in a quasi-judicial manner and catalyse open debates between the various lobbies.

Arising from this analysis, the Commission would consider the advantages and penalties of alternatives for nuclear energy, ranging from rapid growth to exit.

The objective of this work will be to increase understanding of the main factors which will determine the future of nuclear energy and assess realistic policy options. It is hoped that wide dissemination of such information to governments, international bodies, environmental groups, the industry and the public will lead to less confrontation and better individual decisions. What these decisions might be, will vary widely according to the circumstances of individual countries, companies and institutions, but at least there should be better understanding of each others' position if there is a degree of agreement on many of the underlying issues. The Commission will not be asked to produce and 'sell' global solutions.

The structure of the Commission would take the following form:
- The Chair would be a prestigious but neutral figure with wide international experience.
- He/she would be supported by four or five Commissioners, from various parts of the world, also of international standing. At least one should have strong connections with the LDCs.
- There would also be a Steering Committee in a position to give some hands-on advice on the management of the project.
- There would be a Secretariat, possibly of four core members, with the skills necessary to manage the Commission process and prepare briefing papers and reports.

It is vital that the choice of Commissioners and their way of working shows no partiality for or against nuclear energy. Were this not the case, its findings could be ignored as untrustworthy by one side or the other. The Commissioners should be people with wide experience of dealing with major issues and able to appraise complex subjects. Expert advice with varying points of view has to be made available to them.

The mode of operation and work programme would be defined by the Commission. It is envisaged that it would first agree on the main issues it wishes to tackle and request submissions on a series of questions from as wide a range of interests and opinions as possible. These would be considered with the assistance of the staff and of invited experts and, where necessary, debated with the authors of the submissions; public hearings in diverse locations should also be considered, especially if these were to provide a forum for constructive debate between differing factions.

The end-result should be a report setting out the Commission's main findings and a series of meetings with governments, international institutions and the public to disseminate and debate these conclusions.

[i] P. Beck, *Prospects and Strategies for Nuclear Power;* The Royal Institute of International Affairs & Earthscan Publications Ltd., London 1994
[ii] J.G. Morone & E.J. Woodhouse, *The Demise of Nuclear Energy? Lessons for Democratic Control of Technology,* Yale University Press, New Haven CT, 1989; Chapters 3 and 4.
[iii] C. E, Till and Shozaburo Takeda 'Energy over the Centuries: The IFR Option' Chs. 14 and 15. in R. L. .Garwin et al. (eds.), *Managing the Plutonium Surplus: Applications and Technical Options,* NATO ASI Series, Kluwer Academic Publishers, the Netherlands.
[iv] H. J. Otway '*Understanding Public Attitudes towards Nuclear Energy*', paper presented to the IAEA International Conference on 'Nuclear Power and the Public', Geneva 1977.
[v] See for example, M Granger Morgan, '*What would it take to Revitalise Nuclear Power in the US?*' 'Environment', March 1993 p.2.
[vi] C. D. Bowman '*Weapons and Commercial Plutonium Choices - destroy "completely" or store "for ever"* Garvin et al. (eds.) op .cit., Ch.17.

KEY FUTURE ISSUES IN NUCLEAR POWER

John J. Taylor, Vice President Emeritus,

Electric Power Research Institute
3412 Hillview Avenue
Palo Alto, CA 94025

AN ILLUMINATED VISION

During these two days we have heard re-affirmed, among expressions of concern with present problems, the vision that was drawn at the birth of nuclear technology by Enrico Fermi and his associates---that the uranium-plutonium fuel cycle can provide many centuries of the abundant energy needed by a growing and hungry world. But fifty years of experience have illuminated this vision in two important ways:

- Abundant, near limitless resources are necessary, but not sufficient. If that resource cannot be utilized economically, its abundance is academic.
- Developing a means of assuring that commercial fissile material won't be diverted for weapons purposes is a technical and political challenge at least equal to that of developing the power generation system itself.

This session is directed at the issues associated with this illuminated vision:

- The global need for nuclear power, presented by Bernard Appell of the Electricite de France.

- The prospect of sustaining the present generation of light water reactors (LWRs), a necessary path to the need for expanded nuclear fuel resources and a path to establishing public and political confidence in nuclear power, presented by Louis Long of the Southern Nuclear Operating Company.
- The disposition of weapons plutonium, including an option to burn it in nuclear power plants to demonstrate the ability of those plants to reduce nuclear armaments and to increase resistance to nuclear weapons proliferation, presented by Gordon Michaels of the Oak Ridge National Laboratory.

Technology for Global Economic and Environmental Survival and Prosperity
Edited by Kursunoglu *et al.*, Plenum Press, New York, 1997

- The prospects and timing of implementation of the nuclear fuel cycle, presented by Edward Rodwell of the Electric Power Research Institute.
- The international regulatory environment within which the nuclear vision will be nurtured.

PRESERVING THE NUCLEAR POWER OPTION

Mr. Long will concentrate on the efforts being made by the Southern Company to assure that its nuclear power generating capacity, one of the largest in the U. S., will maintain its economic competitiveness in the under rate de-regulation. But there is another aspect of this overall issue which I would like to touch upon briefly: to open the option to expand nuclear power generation in the United States. Peter Beck drew the conclusion this morning that the first priority should be given to preserving this option. The U. S. utilities agree with this conclusion and are further convinced that LWRs are the vehicle for the next step in expansion because of their immense construction and operating experience. The U. S. utilities, because of this conviction, sponsored the Electric Power Research Institute (EPRI) over a decade ago to undertake a program to develop advanced LWRs for that purpose that are safer, more reliable, and more economic. They wanted the option to choose either pressurized water reactors (PWRs) or boiling water reactors (BWRs) and at power outputs of 1300 Mwe or 600 MWe.

The ALWR Program was started in the early 1980s as a purely technical program. After a few years, the program was reviewed by a group of CEO-level industry executives, the Nuclear Power Oversight Committee, to ascertain their continuing interest and support. They were highly complementary as to the technical progress that had been made in defining future LWR plants that would be safer, more reliable, more robust, and more economic than current plants. But they expressed the strong opinion that, even if the technical program was totally successful, the option would not be open in the US unless key institutional conditions were met:

- public and political acceptance
- predictable, stable safety regulation
- predictable economic regulation
- private investment feasibility
- effective radioactive waste management

Accordingly, the utilties commissioned an Industry Strategic Plan[1] which would incorporate, in addition to the already defined technical goals and strategy, the goals and strategies needed to meet the above institutional conditions. That plan was developed and publicized throughout industry and government and even received a favorable editorial from the New York Times.

As the program matured, substantial financial support was provided by the

Department of Energy (DOE) and many international utilities. The Nuclear Regulatory Commission (NRC) participated with detailed technical reviews for the purpose of implementing their standardization policy to achieve a predictable licensing process. Although it has taken longer than had been projected, there have been major technical accomplishments and the overall goals of the Program are nearing completion. Key results to date are:

- Owner/operator technical requirements for future ALWRs, have been defined in-depth in three volumes, and have received Final Design Approval (FDA) from NRC. These requirements substantially exceed the safety goals established by the NRC and, in addition, establish reliability and cost requirements.
- Two 1300 Mwe ALWR designs, the GE ABWR and ABB-CE System 80+, have received FDAs from the NRC. A twin-unit Japanese version of the GE design has been built and is operating in Japan and the System 80+ design is under construction in Korea.
- Two 600 Mwe PWR and BWR designs with passive emergency cooling features have been developed and their passive safety features confirmed by extensive experimentation.
- The NRC review of the Westinghouse PWR passive design is well along, with an FDA expected in the coming year.
- The NRC review of the GE BWR passive design has been shelved pending development by GE of a higher power output version.
- GE has won a contract from Taipower to build the US version of the GE ABWR design in the Republic of China, based on the utility requirements and the first of a kind engineering developed in the ALWR Program.

By contrast, progress in meeting the US institutional conditions has been disappointing. Political acceptance has deteriorated since the passage of the Energy Policy Act of 1992. In spite of the outstanding technical progress, DOE has drastically cut its budget proposals for the ALWR Program and amendments were narrowly defeated last August in the House and Senate to discontinue all funding of the Program. Public acceptance has improved only modestly since the low points of TMI and Chernobyl. The detailed definition of the design certification rule making process leaves significant uncertainties as to whether a stable, predictable licensing process has been achieved. The definition of the economic regulation and private investment issues is under radical change with the de-regulation of the industry, which requires that these issues be completely re-addressed.

In sum, there has been substantial technical progress and limited institutional progress, but the ALWR Program is still underway. It is not only the highest priority nuclear energy supply R&D effort in the U. S., it is the only one. The international support being given this program is of key importance in helping to establish, from an industrial perspective, safety, reliability, and cost goals for future LWR plants.

THE INTERNATIONAL REGULATORY ENVIRONMENT

John J. Taylor, Vice President Emeritus,

Electric Power Research Institute
3412 Hillview Avenue
Palo Alto, CA 94025

INTRODUCTION

It is a widely held premise that a sustainable future for nuclear power depends upon world-wide compliance with consistent regulatory safety standards. Competent, objective national regulatory authorities are needed to establish those standards and to assure that compliance. But another widely held premise is that, since competitive markets result in major economies to the consumer, price regulation of electricity, including nuclear power generation, should give way to competitive pricing.

Competitive markets have often fostered higher quality and safer products as has been shown dramatically in the international competition between Japan and the United States. Yet, the dominance of initial price in the buyers' priority scale does not assure the consistency and uniformity of safety performance needed for nuclear power, because the public perception of its safety is measured by its poorest individual performer.

These potentially conflicting premises strongly suggest the need for safety regulation that requires adequate safety for all nuclear plants so that price competition is eliminated from that portion of the scope of supply. On the other hand, regulation beyond that level of adequacy must be sufficiently cost conscious that it doesn't improperly disadvantage nuclear power in a competitive electricity market. This paper summarizes the present international nuclear power regulatory environment and trends in the context of this regulatory balance.

Technology for Global Economic and Environmental Survival and Prosperity
Edited by Kursunoglu *et al.*, Plenum Press, New York, 1997

285

THE ROLE OF THE INTERNATIONAL ATOMIC ENERGY AGENCY

The International Atomic Energy Agency (IAEA) has provided a forum to pursue international harmonization of nuclear power safety through several key activities: defining basic safety principles, providing generic safety guidance, assisting in safety assurance, and negotiating major international safety protocols and conventions.

Safety Principles

The International Nuclear Safety Advisory Group (INSAG) to the Director General has developed a set of safety principles[1] which have been widely accepted throughout the international nuclear power industry. An up-grade of this document is presently in preparation. Additional INSAG reports have been issued to amplify the principles with respect to safety culture[2], probabilistic risk assessment[3], the application of the safety principles to plants already built to earlier standards[4], potential radiation exposure from accidents[5], and safety through defense-in-depth[6].

Safety Standards

Generic safety standards have been, and are continuing to be, defined by IAEA member expert working groups which provide more detail than the basic principles. The standards are published in the IAEA Safety Series of reports which are categorized hierarchically as "Safety Fundamentals" containing basic objectives, concepts, and principles to ensure safety, "Safety Standards" covering basic requirements to be satisfied to ensure safety for particular activities or application areas, "Safety Guides" containing recommendations relating to the fulfillment of basic requirements, and "Safety Practices" giving practical examples and detailed methods which can be used for the application of "Safety Standards" or "Safety Guides".

These activities must be monitored carefully by the member states to assure that the IAEA activities are not duplicative and that the resulting safety guidance doesn't carry into detailed regulatory requirements. Although some see a potential IAEA role as an international regulator, such a move would create disruptive questions of national sovereignty, would inappropriately ease the burden of safety responsibility from the member states, and would increase regulatory instability. In recognition of this need, the IAEA has recently re-organized the infrastructure of its safety guidance committees to eliminate overlapping responsibilities and to establish an oversight committee to review the activities and output of the various safety guidance committees.

Safety Assistance

The IAEA staff, assisted by volunteer experts, have provided assistance to member states in nuclear plant operational safety matters. Seven plant specific assistance services, or "safety review missions", are available at the request of member states:

(1) ASSET (Assessment of Safety Significant Event Teams) reviews operational safety experience, identifying root causes of incidents, generic safety lessons learned , and the appropriateness of corrective actions.

(2) OSART (Operational Safety Review Teams) assists in the enhancement of safe operation by reviewing operational safety practices and performance, identifying strengths and weaknesses, and giving advice for improvement.

3) ASCOT (Assessment of Safety Culture in Organizations Teams) reviews the effectiveness of safety culture in an organization on the basis of the principles and recommendations of INSAG-4.[2]

(5) Seismic Safety Missions advises on the seismic safety of nuclear power plant sites, as well as plant structures, equipment, and distribution systems.

(6) IPERS (International Peer Review Service) provides an independent review of the probabilistic safety assessment methods applied, the completeness of the assessment, the appropriateness of the input data, and the validity of the results.

(7) IRRT (International Regulatory Review Team) provides advice and assistance to strengthen and enhance the effectiveness of the nuclear regulatory body of a member state.

Following the Chernobyl accident, an IAEA Extrabudgetary Program was launched to provide a much greater level of assistance on the safety of the Soviet-designed water-cooled, graphite moderated channel reactors and pressurized water reactors. Enhanced safety assistance efforts on these systems was also initiated by the European Commission, the World Association of Nuclear Operators (WANO), the Nuclear Energy Agency, the European Bank for Reconstruction and Development, and individual governmental agencies in Asia, Europe, and the U. S.

THE NUCLEAR SAFETY CONVENTION

Following the Chernobyl accident, IAEA took the initiative to establish international protocols to provide for early warning and prompt reporting of nuclear accidents and to foster coordinated international assistance in the event of a serious accident. More recently, IAEA has started the negotiation of nuclear safety conventions to assure continued safety in the peaceful utilization of nuclear energy. The initial convention, the "Convention on Nuclear Safety" was approved by the member states representatives in 1995.

The purpose of this first convention is to obtain the commitment of those member states with nuclear power installations to establish effective national nuclear power regulatory agencies. The convention provides for periodic review of the status of each member's regulatory authority, utilizing a peer review process and periodic, open international meetings to report on regulatory status. Member states that have no nuclear power installations were a party to the development of the convention and will participate in the periodic meetings, reflecting the fact that many of them can be affected if an accident were to occur in a neighboring country.

The U. S. Senate has not yet ratified the convention, which is a legal requirement for all international treaties. The senate has initiated a review of the Convention, the major thrust of which is to assess the cost-effectiveness of the process, its chances of success, and the degree to which it could distract the U. S. Nuclear Regulatory Commission (NRC) from its fundamental regulatory purpose in the U. S. It is hoped that the ratification will precede the first international meeting which is scheduled for April 1997.

It is planned to expand the scope of this initial convention to cover all aspects of the peaceful use of nuclear energy. Discussions are presently underway to extend the convention to cover radioactive waste management. This is not only an important step for nuclear power countries but applies to most member states who need to handle radioactive wastes from medical activities and industrial applications outside of nuclear power production. Pursuit of this ambitious agenda will require focus on the development of national nuclear power regulatory capability. A bi-polar approach is indicated:

Firstly, and the primary motivation of the Nuclear Safety Convention, is building up and broadening the regulatory capability of countries that, for either policy of resource reasons, have not seen fit to provide effective safety regulation of nuclear energy. A spin-off benefit of this build up of capability is that it will establish models of effective regulatory functions that can be readily adapted by countries who wish to introduce nuclear power for the first time. Secondly, improvement is needed in those countries which have well developed regulatory capability, which takes the opposite direction of focusing and streamlining the regulatory processes. There is a need to re-focus the regulatory scope in these countries on the safety of the public by utilizing risk based methodology and moving away from regulatory actions that are not safety beneficial but rather aimed at improving reliability, protecting plant investment, and promoting related economic goals which should be left to the industry to meet.

THE OWNER-OPERATOR SAFETY ROLE

Of particular significance in a discussion of the international regulatory environment is the the definition of safety responsibilities. In its "Basic Safety

Principles"[1], INSAG emphasizes the plant owner-operators' responsibility. Quoting from the document, "The ultimate responsibility for the safety of a nuclear power plant rests with the operating organization. This is in no way diluted by the separate activities and responsibilities of designers, suppliers, constructors and regulators." This responsibility is reinforced by establishing a "safety culture" in the operating organizations as well as all supporting organizations. By "safety culture" is meant "the personal dedication and accountability of all individuals engaged in any activity which has a bearing on the safety of nuclear power plants." This responsibility is further highlighted by the role the industry must play in implementing the "defense-in-depth" concept which provides multiple levels of protection to both the public and the workers, in the form of physical barriers and operational implementation of the associated defenses. Each of the multiple physical barriers prevents the release of radioactive materials, but they envelop each other so that if an inner barrier fails, the next outer barrier holds back the radioactive material. The reliability of these barriers is assured by multiple levels of implementation of "defense-in-depth," which can be characterized by a sequence of concentric design and operational defenses against the release of radiation from the plant.

INSAG also defines the regulator's responsibility. "A legally constituted regulatory organization provides governmental licensing, regulation, and surveillance of the operation of nuclear power plants in respect of their safety. ... The regulatory organization acts independently of designers, constructors, and operators to the extent necessary to ensure that safety is the only mission of the regulatory personnel."

Thus, although the role of the regulator is an essential one, and is a primary focus of this paper, assurance of nuclear plant safety rests fundamentally with the owner-operator of the plant. It is recognition of this basic responsibility which has fostered the development of utility (owner-operator) requirements for new nuclear plants which has been pursued by industry in the U. S., Europe, and Japan.[7]

The EPRI ALWR Program, reviewed briefly in the introduction to this session, has made major progress in formulating such utility requirements. The requirements have been extensively documented[8] and have been reviewed and approved by the U. S. Nuclear Regulatory Commission (NRC). They set safety objectives that are substantially higher than achieved by present plants and that are required by the NRC. Consistent with the responsibilites for safety defined above, the NRC has encouraged these higher safety objectives as an industry initiative and is refraining from incorporating them into new regulations since NRC's present regulations meet its safety goals. This will allow for increased operating flexibility and higher reliability. The requirements also provide for substantially greater thermal margin to reduce the burdens on both the operators and the plant equipment.

Similar work is underway in Europe[9] and Japan[10]. Continuing effort is placed on comparing and reconciling the requirements being developed in Europe and Japan with those developed in the U.S., particularly in the safety area.

CONCLUSION

The importance of sharing safety information and operating on the basis of common standards was dramatically highlighted by the Three Mile Island and Chernobyl accidents and has become an accepted way of life among the owner-operators of nuclear plants and their regulators. The international community is working diligently, aided by international protocols and conventions to become "one nuclear world" from the viewpoint of safety, in recognition that a nuclear accident somewhere becomes a nuclear accident everywhere.

REFERENCES

[1] International Nuclear Safety Advisory Group, *Basic Safety Principles for Nuclear Power Plants*, Series No. 75-INSAG 3, , Vienna, (1991).

[2] International Nuclear Safety Advisory Group, *Safety Culture*, Series No. 75-INSAG 4, IAEA, Vienna, (1991).

[3] International Nuclear Safety Advisory Group, *Probabilistic Safety Assessment*, Series No. 75-INSAG 6, IAEA, Vienna, (1992).

[4] International Nuclear Safety Advisory Group, *A Common Basis for Judging the Safety of Nuclear Power Plants Built to Earlier Standards*, Series No. 75-INSAG 8, IAEA, Vienna, (1995).

[5] International Nuclear Safety Advisory Group, *Potential Exposure in Nuclear Safety*, Series No. 75-INSAG 9, IAEA, Vienna, (1995).

[6] International Nuclear Safety Advisory Group, *Defense in Depth in Nuclear Safety*, Series No. 75-INSAG 10, IAEA, Vienna, (1996).

[7] P. Bacher, K. Iida, and J. J. Taylor, *Forging an International Industrial Consensus on Safety Standards for Future Light Water Reactor Nuclear Power Plants*, World Energy Council 16th Congress, Tokyo (1995).

[8] Electric Power Research Institute, *Advanced Light Water Utility Requirements Document, Vol. 1, ALWR Policy and Summary of Top-Tier Requirements*, EPRI, Palo Alto, CA (1995).

[9] *European Light Water Reactor Utility Requirements*, EDF, Paris, (1994)

[10] *The Japanese Utilities Requirements of Next Century BWR*, ICONE-3; *Development of Next Generation PWR in Japan*, ICONE-3, (1995).

THE POST-ELECTION NUCLEAR AGENDA:

A PENDING DISASTER?

Bertram Wolfe

The American Nuclear Society (ANS) has asked me to discuss the U.S. "Post Election Nuclear Agenda" at their annual meeting in Washington DC next week. As a past ANS president I feel obliged to respond, but it bothers me to do so. For although the Agenda is clear, it is one which will hurt our nation and, in particular, the future welfare of our children and grandchildren. And it indicates a failure of the nuclear industry to effectively inform the public, the media, and our national leaders.

The Post Election Nuclear Agenda is to phase out nuclear power in the United States.

This is not happening in other nations. Early next year I will be attending a celebration in Japan to commemorate the commercial operation of their newest nuclear plant; one designed and constructed by a U.S.-Japan team led by General Electric, and built in four years. The Japanese plan to build many more of these plants in the coming years, and there are similar plans to expand nuclear power in other countries around the world.

As noted, the Japanese nuclear plant, built to U.S. technical and safety standards, completed construction in four years. But in this country, due to bureaucratic government impediments, lack of government responsibility, and court delays initiated by anti-nuclear groups, it takes a dozen or more years to build a nuclear plant. This leads to costs which makes nuclear power plants uneconomic in the United States. Similarly, our nuclear waste programs have been impeded by anti-nuclear forces who use politics and our courts to cause delays. They then scare the public by arguing that there is no way to get rid of nuclear wastes.

The government has under development a new nuclear plant licensing procedure called "standardized licensing", intended to improve the U.S. situation, but it has no plans to demonstrate that the new licensing system works. Under these circumstances, who would take the risk of committing to a new multi-billion dollar nuclear project in this country? The Department of Energy now projects that in the next 20 years a third of the nuclear plants in this country will be shut down, and not replaced, as their licenses expire.

The post election nuclear agenda is clear, but unlike the past twenty years when

Technology for Global Economic and Environmental Survival and Prosperity
Edited by Kursunoglu *et al.*, Plenum Press, New York, 1997

the U.S. has had a surplus of electrical generation capacity, we will need additional capacity in the future. More crucial is the world situation, where in the first half of the next century the World Bank and the United Nations project that the third world population of four billion people will double, leading to a total world population of 10 billion people by 2050. Even with strong conservation, this will lead to a tripling of world energy needs.

If fossil fuels are used to meet the next century's expanding energy needs, we face a potential environmental disaster from atmospheric pollution and the greenhouse effect (global warming). And we face potential wars over scarce energy supplies. Consider for example, why we fought Iraq in Kuwait, and why we have stationed our soldiers in Saudi Arabia, where a number were killed, and the others are still at risk. Indeed, concern over energy supplies was one reason Japan entered WWII.

Hopefully, solar or wind power, or fusion or cold fusion will turn out to be practical and we should try to develop them; but they have such basic problems that, in my view, *it is irresponsible to base our future on them.* The only demonstrated safe, clean, energy source which can provide needed energy for as long as "people inherit the earth" is nuclear energy.

I made many of the points above in a paper entitled "Why Environmentalists Should Support Nuclear Power" which was published by the National Academy of Sciences in the summer edition of "Issues in Science and Technology." I sent a copy to Vice President Gore. I am not sure that he personally read my paper; but nevertheless he was kind enough to respond, and he said: "...we believe strongly that this country can meet its future energy demands by incorporating better efforts toward conservation and the widespread use of alternative energy. To that end, this Administration is opposed to increased reliance on nuclear power."

Thus, as noted, the post election nuclear agenda is clear. It is part of the "Bridge to the Future" which President Clinton and Vice President Gore are building. There will likely be no major observable effects during the four years of this Administration since, although we are failing to meet the President's greenhouse gas reduction targets, there are adequate fossil fuel energy supplies - including our fuel imports. But do our leaders understand the dangerous terrain to which their Bridge will lead our nation in the future?

The U.S. has been a leader in developing nuclear energy around the world, and has been a major influence in setting both high safety standards, and arrangements to help prevent peaceful nuclear power from being used to develop nuclear weapons. Not a single person has been harmed by operation of nuclear plants built to U.S. standards - including Three Mile Island. Indeed, one might note that nuclear power is saving lives by reducing fossil fuel emissions which, in the U.S. alone, are estimated to cause tens of thousands of premature deaths each year. This yearly number is more than the total death toll from Chernobyl, a reactor which did not meet U.S. standards, and would not have been permitted to be built here.

It should be understood that all technologies have both problems and benefits. Despite their environmental and health problems, fossil fuels overall, are saving US lives by providing needed energy to allow us to live decently relative to those in most of the world who live in poverty, and only have a fraction of the energy which is available to each of us. But, as indicated, the coming decades of increased world energy use will put strains on fossil fuel supplies and the environment of the world, and there seems no available solution to a potential world disaster without a major expansion of the use of nuclear energy.

The U.S. nuclear industry can be proud that, despite our domestic problems, we have been able to maintain our technical competence and nuclear leadership abroad; as indicated by our recent winning of an international competition to provide two new nuclear plants to Taiwan. But, on our present course we are in danger of losing our leadership.

For example, Chang Kan Lee, an energy specialist from Korea, has stated: "...The U.S. has been a godfather in the nuclear sector. In the past two decades, however, this nuclear godfather has not been very active with nuclear projects, except in the area of nuclear regulation. So the western pacific beneficiaries now have to hold the nuclear candle themselves . . . and time may come when the former god-father will come to his god-sons for help."

Perhaps our children should be thankful that, if the post election nuclear agenda proceeds, in the future they may be able to rely on overseas nuclear technology to revive our needed nuclear capability. But we should recognize the international risks we will face if we lose our world nuclear influence and leadership; and the problems we will face in reviving nuclear energy here, if we lose our technical competence. One may note, for example, that many of our nuclear experts are retiring, and the number of students in our university nuclear engineering departments is rapidly declining.

Let me conclude by noting that we in the nuclear energy field should share part of the blame for the potentially disastrous post election nuclear agenda. Although the U.S. nuclear industry has done an outstanding technical job in meeting President Eisenhower's Atoms for Peace goals, we have not done a good job in educating the public and many of our government leaders, including President Clinton and Vice President Gore. We have let them be mislead by anti-nuclear rhetoric and unsound advice.

Thus, I have sent send them copies of my ANS talk, and offered to set up a personal meeting with them in which those highly knowledgeable about energy and nuclear energy can provide some perspective. If this offer were accepted, I believe it could greatly benefit our nation by providing information to our President and Vice President which may help them modify the construction of their bridge; so that our children and grandchildren will be able to cross over into a safer and more satisfying future.

Dr. Bertram Wolfe is Fellow and Past President of the American Nuclear Society, amember of the National Academy of Engineering, and the recipient of a number of honors for his contributions to the energy and nuclear energy fields. He retired as head of GE's nuclear energy business in 1992, and is now an independent consultant on several Boards of Directors and advisory committees.

INDEX